EVOLUTION
THE GREAT
DEBATE

TO CHRISTINE AND MARY

EVOLUTION THE GREAT DEBATE

Vernon Blackmore and Andrew Page

A LION BOOK
Oxford Batavia Sydney

Published by
Lion Publishing plc
Sandy Lane West, Littlemore, Oxford, England
ISBN 0 7459 1208 7 (cased); 0 7459 1650 3 (paperback)
Lion Publishing Corporation
1705 Hubbard Avenue, Batavia, Illinois 60510, USA
ISBN 0 7459 1208 7 (cased); 0 7459 1650 3 (paperback)
Albatross Books Pty Ltd
PO Box 320, Sutherland, NSW 2232, Australia
ISBN 0 7324 0048 1 (cased); 0 7324 0049 X (paperback)

Printed and bound in Spain

British Library Cataloguing in Publication Data
Blackmore, Vernon
    Evolution.
    1. Man. Evolution. Theories. History
    I. Title   II. Page, Andrew
    573.2'01

    ISBN 0–7459–1208–7

CONTENTS

AUTHORS' NOTE

We would find it impossible to list all the books and articles
which have influenced our thinking and which have in part
shaped this book. But we would like gratefully to
acknowledge:

F.J. Ayala, *Population and Evolutionary Genetics*, Benjamin/
Cummings, 1982; L. Barber, *The Heyday of Natural History*,
Jonathan Cape, 1980; R.W. Clark, *The Survival of Charles
Darwin*, Weidenfeld and Nicolson, 1984; C.C. Gillispie,
Genesis and Geology, Harper and Row, 1951; E. Mayr, *The
Growth of Biological Thought*, Harvard University Press, 1982;
J.R. Moore, *The Post-Darwinian Controversies*, Cambridge
University Press, 1979; D.R. Oldroyd, *Darwinian Impacts*,
Open University Press, 1980; A.R. Peacocke, *Creation and the
World of Science*, Clarendon Press 1979; R.E. Ricklefs, *Ecology*,
Nelson, 1973; G.L. Stebbins, *Processes of Organic Evolution*,
Prentice-Hall, 1966.

INTRODUCTION

The scientist and author Isaac Asimov is a worried man. The rise of 'creation science' in the United States appears to him to threaten the whole edifice of science. Creationists believe that God made the world in just six days, as a literal reading of the opening chapter of the Bible demands. Asimov writes:

> 'With creation in the saddle American science will wither. We will raise a generation of ignoramuses. We will inevitably recede into the backwaters of civilization.'

But his opponents are not, anyway, impressed with our late twentieth-century civilization. Darwin's evolution, they claim, has sapped our moral strength, and made us question the existence of God. America is no longer beautiful. And the poisonous ivy is evolution which 'is at its roots,' according to the creationist leader Henry Morris, 'the anti-God conspiracy of Satan himself.'

This book is not about the rights and wrongs of evolution or creation science. There is no detailed comparing of fossils, calculating of probabilities, or checking the validity of one theory against another. For there is a much more fascinating story to be told: the history of the idea of evolution itself and in its wake the troubled waters of religious argument. And here we will find not just religion, but the flotsam of political and social creeds, and the deep human craving for an understanding of our origins. Here, too, we will meet science. And we will encounter scientists. Hopefully we will learn to tell the difference.

'Which would you prefer to have happened?' quizzed a Christian minister as we concluded a talk on the issues underlying the evolution-creation debate. We had tried not to come down strongly on either side (for our brief was simply to discuss the issues), but the pastor would not let us slip away that easily. He wanted to know which of the two options we preferred.

Prefer is an interesting word. As we pointed out to him, it matters not a jot what we prefer. For someone with religious faith the search is to discover what God has actually done. Has he created by a sudden miracle or by the mechanism of evolution? Do the fossils witness to gradual development or to creation in six days? Our own wishes on the matter are as irrelevant as whether we prefer strawberry jam to blackberry jam.

But by talking of preferences we had entered onto new territory. We had left behind the simple understanding of science as making decisions based on a catalogue of facts. We were acting as human beings who judge the validity of one fact or another according to whether it fits in with our cherished ideas. Perhaps that is a bit harsh and too dismissive of the noble search for scientific truth. But our friend thought that God was somehow more powerful if he worked through sudden miracles rather than by slow evolution. It was as though he needed a bigger God, and so he read his science expecting, and almost hoping for, miracles. For him the fossils confirmed he was right.

On many scientific issues a neutral stance is possible. Unbiased research is expected.

Neither of us would have studied and worked as scientists unless we believed that. But an understanding of our origins is so crucial to the way we view ourselves that in the study of evolution such objectivity is rarely achieved. At least our friend, with his talk of 'preference', was open about the factors which influenced his thinking. More often they lie hidden within our almost subconscious cultural, religious and political views.

This blindness is not limited to the religious. In recent years new developments in understanding evolution have been labelled (and so dismissed) as Marxist. For a long time humanists have used the theory as a stick with which to chase God out of his universe, and then have used it as the framework for their own philosophical ideas. In the last century ideas of evolution were first rejected as socially disruptive (they spoke of change), only to be later endorsed as the justification for ruthless business competition and development.

Most of us have grown up to think that Darwin disproved the Bible. On that score this book, too, is biased. We write it out of the personal experience and conviction that science and religious faith are not incompatible. There is an old African proverb which says, 'Until the lions have their historians, tales of hunting will always glorify the hunter.' Well, the history of science has often been presented as the victory of the scientific way of picturing the world over outdated religious mythology. But there is another story to tell.

Historians of science ponder why the scientific spirit arose in Western society precisely at the time it did. Why not 200 years earlier, or why not in China? And they also ask why Darwin's theory of evolution was not discovered centuries earlier, by Newton or Galileo. In their answers they invariably highlight the general philosophy or economics of the times, and it is these 'outside-of-science' matters which bear such an influence. It is these factors which make the history of science so interesting. The story is not simply of one theory becoming established, only to be disproved and give way to another. It is the story of people with varying beliefs, pursuing truth for its own sake but also muddying the waters of knowledge with their own prejudices and preferences.

Yes, there is another story to tell.

1
THE GREAT SYSTEM
OF NATURE

Charles Darwin published his famous *Origin of Species* in 1859. He tried to explain how, in the long history of the earth, the present animals and plants had arisen from simpler organisms. It was a revolutionary picture of gradual evolution over endless stretches of time. In so doing he shocked Victorian religious sensibilities for he appeared to be saying that we are descended from apes rather than created by God.

The real story is more complex, for some scientists saw Darwin's work as exalting God rather than destroying faith. But before we can plunge into these controversies we need first to examine the views leading up to Darwin.

He did not write in an intellectual vacuum, and we should note some of the biological concepts used in earlier centuries before we can see how Darwin brought in his new age of biology. In particular, naturalists before Darwin stressed that species could not change and that the purpose of their study was to classify the flora and fauna. But gradually they realized that species were not forever the same, and that fossils were incredibly old. Whole species had

We call cheetahs 'big cats', rightly since they are members of the cat family. From the days of Aristotle onwards, organisms were classified in families.

died out; new forms had been discovered. How could this be so, they asked, if the Bible said everything was created in six days?

The Greeks

'I cannot understand why you scientific people make such a fuss about Darwin. Why, it's all in Lucretius!' So wrote Matthew Arnold to John Judd in 1871.

We must not go away with the idea that before Charles Darwin no one had ever thought of evolution. The concept dates back more than two millennia, to those first philosophers, the Greeks.

The centre for the earliest Greek philosopher-scientists was not Athens but Miletus, the Ionian trade centre on the coast of the Aegean. Here they tried to give a general account of the world, of how it is made, and how it changes. Their various theories need not detain us, but one school of thought understood the world to be composed of small indestructible components. When things change it is but a re-ordering of these fundamental (and unchanging) particles. In Greek 'a-tom' means 'uncuttable', and these philosophers were known as 'atomists'. They are best known to us through the writings of Lucretius, a Roman philosopher of the first century BC.

Change was at the basis of their understanding. And how particularly did plants and animals evolve? Thales, one of these Ionian philosophers writing around 550BC, certainly believed that all life arose initially from marine or aquatic beginnings — an idea still held today. Anaximander believed in evolution, both in the cosmos and in the natural world. Living creatures arose from the 'moist element' as it was evaporated by the sun. Mankind, like every other animal, was descended from fishes.

Empedocles, from the south coast of Sicily, wrote:

'The first generation of animals and plants were not complete but consisted of separate limbs not joined together; the second arising from the joining of these limbs were creatures in dreams; the third was the generation of whole mature forms. Here sprang up many faces without necks, arms without shoulders, unattached, and eyes stayed alone in need of foreheads. But as one divine element mingled with another, these things fell together as each chanced to meet others, and many other things besides were constantly resulting. Whenever, then, everything turned out as it would have if it were happening for a purpose, there the creatures survived, being accidentally compounded in a suitable way; but where this did not happen, the creatures perished and are perishing still.'

Fanciful ideas of this type were dreamed up and stated with little reference to the real world. Yet, if you prune away the fantastic, you are left with the ideas of evolution, perhaps even of natural selection — the evolutionary mechanism proposed by Darwin himself some 2,300 years later!

Darwin, in the historical sketch which introduces his epoch-making *Origin of Species* (1859), passes over allusions to evolution in the classical writers, save for a partial quotation of the above found in one of Aristotle's writings. Instead he hurries on to the eighteenth-century French scientist Lamarck. But it is right for us to dwell a little longer on the Greeks, for they bequeathed to biological science some important concepts for classifying the natural world.

Aristotle and the zoo

It is obvious that a cat is more akin to a leopard than to a human being. Today we talk in terms of the 'cat family', which embraces not only the leopard but also the lion, tiger and cheetah. The great philosopher Aristotle, born in 384BC, pupil of Plato and tutor to Alexander the Great, was a keen observer of the world around him. He sought, with the help of his pupils, to catalogue all human knowledge. In the field of natural history his work was excellent, and he grouped organisms into family units based on commonly observed similarities. He saw monkeys as an intermediate form between humanity, the bipeds, and the other animals, which were four-footed. He made divisions between birds, fishes and whales.

Aristotle's lasting legacy to biology was twofold:

☐ *He placed importance on the form or structure (that is, the morphology) of the organism in his classification system.* He also made classifications according to how the animals lived. 'Animals can be grouped according to their modes of life, their activities, their habits and the parts of their bodies.'

☐ *His classification of organisms expressed degrees of 'perfection'.* To Aristotle it was obvious that the lower orders of nature (the inanimate plants) were not as advanced as the mammals. And if you compared one organism with another, right across the spectrum of nature, you saw a sort of ladder or scale of perfection. This idea was to prove important in later centuries when naturalists realized that, over a period of time, such a scale could represent development of complexity with time, with organisms progressing from rung to rung as they increased in 'perfection'.

But Aristotle never thought of organisms clambering up the ladder of complexity. His focus was on classifying the natural world, not on tracing its development with time. The scale of perfection only expressed what he saw; it was not a guide to its history.

If we visit a large art gallery we may view the paintings grouped according to their style. We may prefer a Pissarro to a Picasso. An art historian will quickly point out how over the centuries one style has developed into another, and even show transitional pictures which retain some of the old style while reflecting the new. But, if we choose, we can wander around the gallery quite oblivious of this knowledge. All we know is the pictures we like, and that all the

Aristotle, born 348BC, is known chiefly as a philosopher. But his thought ranged widely, and he was interested in the whole great system of the natural world.

impressionists are kept together in one area of the gallery.

In the same way Aristotle did not think in terms of historical connections. The boundaries between one group of animals and the next were distinct and each had its place in the collection. There was no thought of one family turning into another. Now when we visit a zoo we see the natural world through Aristotelian eyes. We pass from the lizards to the monkey house and then on to the giraffes. Each animal type is confined in its own cage. We are walking through a classification of the natural world, and it is not easy to grasp how one species might be the ancestor of another. The eighteenth-century biologists, the forerunners of Darwin, were great collectors and namers, much as Aristotle had been. They sought to define the present fixed groupings within nature, and seldom speculated on how those groupings arose or whether, given time, they would change. They believed in the *fixity* of species, not in evolution.

The Great Chain of Being

The classifying approach to biology may also be compared to working in a postal sorting office. (With this analogy we begin to move away from

When we visit a zoo, we pass from house to house seeing distinct types of animal in each. Early biologists approached the natural world in a similar way, naming and classifying natural life.

Aristotle.) True, each district has its own pigeon hole, and each letter must be placed in its right box. But if you look at a street map of the area covered by the sorting office a different, less cut-and-dried picture appears. Districts merge into each other: cross a street and the district changes from X1 to X17. The boundaries are often man-made for our convenience. And in future years a new postal system may mean the boundaries being redrawn.

Or, to use another picture, when a do-it-yourself enthusiast buys his paint the labelled tins are clearly grouped together according to each tint and hue. The colours on the paint selector chart may have completely different names, but there are often only subtle changes between one colour and the next. For some purposes (such as posting a letter or ordering a tin of paint) you will need to define distinct categories. But at other times (such as when you walk your dog or mix your paint) it is clear how arbitrary such boundaries are.

And so in biology there is a parallel tradition to the rigid naming approach. Rather than stressing the distinctiveness of each species, this approach emphasizes the continuous spread of animal and plant diversity. It is as though the rungs of Aristotle's ladder are so close together they form a slide. Listen to Plotinus, a philosopher of the third century AD:

> 'The attentive observer will discover a connection of parts, from the Supreme God down to the last dregs of things, mutually linked together and without a break. And this is like Homer's golden chain, which God, he says, bade hang down from heaven to earth.'

This 'Great Chain of Being', representing an infinity of forms, was a persistent picture of nature in subsequent centuries. For Christian philosophers it represented the fulness of God's creation: on the spectrum from simple organisms to mankind there were no gaps. John Wesley, who was a popularizer of science in the eighteenth century, gives us an example of this, in his *Survey of the Wisdom of God in Creation*:

'The whole Progress of Nature is so gradual, that the entire Chasm from Plant to Man is filled up with divers Kinds of Creatures, rising one above another by so gentle an Ascent that the Transitions from one Species to another are almost insensible.'

The term 'missing link' was used in the context of the Great Chain of Being many centuries before our modern search for humanoid ancestors. It represented an apparent gap in this chain, which was believed to be continuous. Further explorations to yet uncharted corners of the world would reveal these 'links', and so restore humanity's confidence in the plenitude of God's created works.

To the modern reader, brought up in a world familiar with change and development, it will seem surprising that this chain or spectrum of being did not automatically lead to ideas of animal evolution. But for the scientists of the day it was simply a way of understanding creation as it stood before them at that moment. After all, the do-it-yourself painter is content to survey the colour chart and the stock of paint tins available. Unless he mixes his own paint he is unlikely to speculate on how the various shades are all derived from the primary colours! And the classifying naturalists were something the same.

So, without even considering evolution, two differing questions were asked about the natural world:

☐ *Are there distinct kinds in nature?*

☐ *Or are divisions into families and species man-made?*

Should we, like Aristotle, picture nature as a zoo? Or is the truth nearer that of a spectrum of forms? These were questions which returned in the years prior to Darwin.

However, Aristotle's work lay largely unknown for many centuries. It was only in the thirteenth century that it filtered through to European scholars via Arab sources. With the scientific renaissance between 1450 and 1650

there came a renewed interest in nature. We all know the story of Galileo and the church, and so we imagine that the church persecuted rather than patronized the early scientists, but the facts are otherwise. Christians believe that God is Creator of the world, and this belief gave impetus to investigations of nature. Science was seen as 'thinking God's thoughts after him', and the natural order of the world was a reflection of God's activity.

This period of history was also the time of great explorations. From the fourteenth century onwards travellers, many of them Spanish or

Christopher Columbus' ship, the Santa Maria, was one of many to go on great journeys of discovery. In the new lands the explorers found new kinds of animal and plant. The variety of nature was proving more prolific than had ever been imagined.

Portuguese, brought back to Europe tales of strange animals and lands. Here true eye-witness reports were liberally mixed with legend. Gradually this surge in knowledge was assimilated, and during the sixteenth century a number of encyclopedic works on natural history were produced. One of these, *Historia Animalium* by the Swiss Konrad von Gesner, ran to 4,500 folio pages and included the famous woodcut of a rhinoceros reputedly by Albrecht Durer. (The beast in question was shipped all the way from Asia to Lisbon.) Throughout his book Gesner based his classification on that of Aristotle, a lasting testament to the Greek philosopher.

The great systematists

The tradition of classification was a long one. It was continued in the seventeenth century by the great English parson-cum-naturalist John Ray (1627-1705). The son of a blacksmith in the Essex village of Black Notley, Ray studied at Cambridge University and was ordained, so adding another name to the long list of clergymen active in natural history through the centuries. His particular interest in botany was encouraged by his mother, who was a herbalist, and so had a considerable knowledge of the local flora.

John Ray, a seventeenth-century English clergyman, took the science of biological classification to new degrees of detail. His reverence for natural life was an integral part of his reverence for God.

At that time the unquestioned belief in the fixity of species made evolutionary thought impossible. Nevertheless, looking back down the centuries we can see how important Ray's studies were. An orderly and classified arrangement of life needed to be made before the historical connections between one species and another could even be recognized. And so Ray set about his work of classifying the natural world with no evolutionary scheme in mind. The different species of living creatures were so many aspects of the Creator's original plan. In fact Ray titled his most famous book, published

Animal illustration was a developed art by the sixteenth century. Albrecht Durer's woodcut of a rhinoceros is one celebrated example.

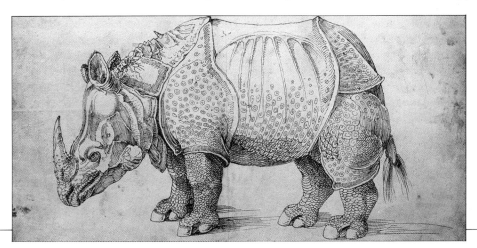

in 1691, *The Wisdom of God Manifested in the Works of Creation*.

> 'What can we infer from all this? If the number of Creatures be so exceeding great, how great, nay, immense, must needs be the Power and Wisdom of him who form'd them all!'

He arranged numerous species of plants into groups according to the structure of their fruits, leaves and flowers, estimating that there were over 18,000 different plant species. He did similar work in zoology where he was the first to realize that some vertebrates use lungs to breathe while others use gills. He was also the first to make a correct classification of bats as mammals rather than as birds, and it is to John Ray, the 'English Aristotle', that we owe the concept of what a species is: a group of animals or plants capable of interbreeding.

The question that was later to vex biologists was, 'Can species change?' Forms might vary within the species. It is only too apparent that we humans differ widely from one another in size and shape. But could species themselves change? Ray believed not.

> 'The number of species in nature is fixed and limited, and as we may reasonably believe, constant and unchangeable from the first creation to the present day.'

But if Ray and his contemporaries believed that nothing in nature had changed, how were they to explain the countless marine fossils found embedded in rocks? The days of collecting dinosaur skeletons were still some way off, but smaller fossils were well known. How did they come to be there? Did they not witness to organisms long since extinct? The Greeks and Romans had (rightly) deduced they were petrified remains and attributed their presence on dry land and even on the summits of mountains to vast inundations of the land by the sea. But since many fossils were of organisms no longer living, this implied that whole species had died out — a direct challenge to Ray's belief that nature was 'unchangeable from the

first . . . ' Alternative explanations would have to be sought.

One such interpretation supposed that fossils were the unfinished work of a *vis plastica* or creative force, capable of changing the inorganic into the organic. This force had given form to the fossil, but not life. Writers of the Renaissance considered these fossils to be 'sports' or jokes of nature, imitations of living forms created to adorn the secret parts of the earth. But Ray realized that such exact similarities to real plants were hardly ascribable to accident. We find him recoiling from the consequences of such thinking:

> 'There follows such a train of consequences, as seem to shock the Scripture-History of ye novity of the World; at least they overthrow the opinion generally received and not without good reason, among Divines and Philosophers, that since ye first Creation there have been no species of Animals or Vegetables lost, no new ones produced.'

Ray died in 1705. Two years later the greatest of systematists was born: the Swede Carl von Linne, better known to us by the Latin version of his name, Linnaeus. Initially committed to an original creation of fixed species Linnaeus came, in time, to echo the questions and the doubts of Ray.

Linnaeus, the second Adam

Linnaeus trained in medicine, but spent the greater part of his working life teaching natural history at the University of Uppsala in Sweden. His father had originally intended that Carl would enter the church, but his school teachers declared him a dunce and fit only for menial jobs. It was botany, and to a lesser extent medicine, which eventually sparked off his latent brilliance. Rather appropriately, 'Linnaeus' derives from the Swedish for a lime tree, and was a surname adopted by his grandfather when he had gone up to university many years before.

In his mid-twenties Linnaeus made a name

for himself by exploring Lapland, a country not then as accessible as today's package tours to the 'land of the midnight sun' have since made it. In the 1730s this was considered a savage country, cut off by its fearsome climate and inhospitable people. Yet Linnaeus returned from this uncharted region with many new species of plant and with considerable enthusiasm for his future work. That work lay in describing and grouping all the then-known plants and animals. Indeed it is to Linnaeus that we owe the hierarchical system of classification used by all biologists today.

Linnaeus took up a suggestion given by Casper Bauhin in the previous century, and identified every organism by two names. The first, generic name, indicated the general grouping: 'doglike', for instance. The second represented the specific species within that genus, such as the wolf. Hence the two-part name, *Canis lupus*: 'dog wolf'. Prior to this, naturalists had tried to give specific names which incorporated all the distinguishing features of the organism. But, as writer Lynn Barber points out, who could hope to remember that *Achillea foliis duplicato-pinnatis glabris laciniis linearibus acute laciniatis* was humble milfoil?

The simplicity of Linnaeus' system and the

HOW MANY KINGDOMS?

Linnaeus classified the natural world using double Latin name tags — a genus name (such as *Homo*) followed by a species name (such as *sapiens*). The different species may be grouped together into a genus because of the similarities between them. Similar genera can then be grouped together into families, similar families into orders, similar orders into classes, similar classes into phyla and similar phyla into kingdoms. Any individual can be placed in this hierarchy. Mankind is given the specific name *Homo sapiens*, which means 'Man the wise', and our full description is:

    Kingdom: Animalia
    Phylum: Vertebrata
    Class: Mammalia
    Order: Primates
    Family: Hominidae
    Genus: Homo
    Species: Sapiens

Since Linnaeus' day there has been something like a ten-fold increase in the number of species known to us. Each new discovery is placed into the hierarchy either in a group already established or, if the organism is sufficiently distinctive to warrant it, in a new group created at some suitable point. To the biologist the system of classification is a tool to be used and, should the need arise, it can be modified to make it more useful. It is not something which is fixed for all time. At present something like 23 phyla, 80 classes and 350 orders of animals are recognized, but there is a constant need to update this structure.

For Linnaeus there were only two possible categories at the level of the kingdom: plant or animal. Such a division still seems reasonable to us today and for most of our common needs it is quite sufficient. But if we begin to look closely at how we characterize a plant or an animal we find individuals which do not fit neatly into our scheme. Plants, we say, have roots in the ground and produce their own food by the interaction of sunlight with a green pigment in their leaves called chlorophyll. Animals, on the other hand, can move around and feed on other animals or on plants. This classification is reasonably successful until we look down a microscope and find the Euglena. The Euglena is a single-celled organism which swims around in ponds and puddles by means of a long whip-like appendage called a flagellum. In this respect it is an animal. But it also contains chlorophyll and makes its own food with the aid of sunlight. In this respect it behaves like a plant. So, which is it?

The blending of the characters of both the plant and animal kingdoms is a feature of a number of single-celled organisms. When Linnaeus was setting up his system of classification, microscopes were not very good and studies of the microscopic world were poorly advanced. By 1866 the inadequacies of the two-kingdom system of classification led the German biologist Ernst Haeckel to propose a third kingdom, the *protista*, into which all micro-organisms were placed.

Under Haeckel's classification, bacteria are single-celled organisms and are included in the protista. But the more we learn about bacteria the more we discover that they are profoundly different from other members of the protista like Euglena. Euglena, for instance, has a nucleus inside it which contains the genetic information for controlling the cell; bacteria do not have this nucleus. This does not mean that bacteria lack a control mechanism. The mechanism exists as a circular strand of genetic material but it is not enclosed in a membrane to form a discrete nucleus. These cells without a true nucleus are too different to be classed alongside Euglena and biologists felt that they

The Swedish naturalist Linnaeus devised the two-name system of classifying species which is still used today.

excellence of his descriptions won the day. Indeed naturalists, both amateur and professional, came to regard it as the highest accolade to have their own name incorporated within one of Linnaeus' Latin terms. He reduced the confused mass of long-winded descriptions to an elegant all-encompassing system; so much

should be placed in a separate kingdom. During the 1930s and 40s Copeland developed a four-kingdom classification in which he added *monera* alongside the existing protista, plants and animals.

In 1969 R.H. Whittaker proposed a further spilt. His classification included a fifth kingdom for the fungi. The fungi obtain their nutrition by absorption, a process characteristically different from either the photosynthetic mechanism of the plants or the ingestion mechanism of the animals. Perhaps in the future there will be further additions, but the classification now rests with these five kingdoms:

☐ *Monera*: contains organisms like bacteria and blue-green algae;
☐ *Protista*: contains mainly one-celled organisms with a true nucleus;
☐ *Plantae*: contains multicellular organisms with rigid walls. The principal method of nutrition is by photosynthesis;
☐ *Fungi*: contains organisms which

have many cells in long, rigid-walled filamentous structures (called mycelia). The sole method of nutrition is by absorption;
☐ *Animalia*: contains organisms composed of many wall-less cells. Nutrition is primarily by ingestion and subsequent internal digestion.

Systems of classification may seem of little relevance to evolution. But the systematic organization of the five kingdoms is intended to reflect the evolutionary descent of the various groups. Specialists may debate the relationships between the groups but there is no doubt that these evolutionary relationships exist. A family photo-

Natural life is rich and varied, almost bewildering in its variety. It was this very profusion which led to naturalists' emphasis on classifying species.

graph which includes uncles, aunts, and cousins, as well as the immediate family, quickly indicates who shares which family characteristics. From the photograph alone a complete stranger might be able to work out the family tree. So in the eighteenth century the attempts to classify the staggering diversity of the living world led to a system of classifying which suggested lines of descent.

so that if an organism was not recorded in Linnaeus' famed *System Naturae* then some doubted whether it really existed!

In developing his classification system Linnaeus supposed he was unearthing the very order of God's creation. In the book of Genesis Adam named the creatures God had made, and Linnaeus saw himself as a second Adam. He was a firm believer in the biblical doctrine of creation. The original Garden of Eden was, so he theorized, a small island somewhere near

Linnaeus built up a vast collection of botanical specimens, of which these leaves are an example.

the equator. He observed that tide levels around Sweden were gradually receding, and supposed that the seas over the whole globe were doing the same. Since the days of creation more and more land had been thus exposed, allowing the original animals to multiply and migrate, creating the situation as in the eighteenth century. Certainly all species had remained constant since the creation:

> 'There are as many species in existence as were brought forth by the Supreme Being in the beginning ... and consequently there cannot be more species now than at the moment of creation.'

However, like Ray before him, he was to revise his opinions in the light of further evidence.

Planting doubts

Linnaeus' change of thinking began in 1741 when a student, Magnus Zioberg, brought to him a specimen of a plant. It showed all the features of the genus *Linaria* except that the flower structure was very different. He called the new strand *Peloria* (the word means monster), and at first thought it was a trick or a hybrid. But hybrids were generally thought to be sterile, like mules, yet this one seemed perfectly fertile. In 1744 he wrote *Treatise on Peloria* in which he gives his opinion:

> 'Nothing could be more miraculous than the story of this plant, in that the deformed offspring of a plant which normally produces irregular flowers, now produces regular ones. This is not only a deviation from its mother genus, but from the whole class; a unique phenomenon in botany.'

As the *Peloria* could propagate he had to acknowledge that: 'It is possible for new species to come into existence within the plant kingdom.'

However, Linnaeus is not remembered for this new departure in outlook. His lasting fame lies in his meticulous classifications. His *System*

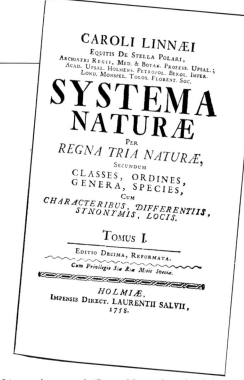

Linnaeus' great work, 'System Naturae', is a detailed catalogue of plants and animals.

Naturae went through twelve successive editions during his lifetime. Linnaeus wrote the first, some fourteen pages, while still a student; the last ran to over 2,000 pages. The overwhelming emphasis throughout was the fixed order of creation. The title page of the tenth and definitive edition (1758), bears the dedication:

> 'O Jehovah how ample are Thy works! How wisely Thou hast fashioned them! How full the earth is of Thy possessions!'

The book is a supreme catalogue of those divine possessions. Because he saw the possibility of new species arising through crossbreeding, he cautiously removed from later editions of his book the statement that no new species can arise. Yet his system is Aristotelian through and through: each animal grouping is distinct and there is no concession to natural mutability of species with time. Indeed Linnaeus'

SIEGESBECKIA — A STINKING WEED

The classification of plants by Linnaeus caused some offence.

He demonstrated conclusively that the reproduction of plants had a sexual basis, with pollination being accomplished by the male and female organs, called stamens and pistils respectively. Accordingly he divided all flowering plants into classes based on their male sexual organs, with subsequent divisions into orders based on the female organs. Biologically this made sense, but in an age sensitive to sexual language it proved repugnant, especially since Linnaeus went on to describe classes such as *Diandria* as 'Two husbands in the same marriage' and *Polyandria* as 'Twenty males or more in the same bed with the female'!

'God never would,' wrote Johann Siegesbeck, professor of St Petersberg, 'have allowed such odious vice as that several males (anthers) should possess one wife (pistil) in common, or that a true husband should, in certain composite flowers, besides its legitimate partner, have near it illegitimate mistresses.'

Linnaeus retaliated to this supposed 'loathsome harlotry' by giving the name Siegesbeckia to a particularly stinking weed!

Throughout the course of his life Linnaeus formed a vast collection of specimens, not only of plants but also of insects, shells and mineral specimens. In the end he was obliged to build a museum to hold his massive hoard, and on his death in 1778 his wife put the lot up for sale. Much to the chagrin of the Swedish government the purchaser was an Englishman, Sir Joseph Banks. Story has it that a Swedish warship tried to stop the English sailing away with the booty. Be that as it may, the treasure

English naturalist Sir Joseph Banks bought Linnaeus' collection of specimens. It is still held in the Banks Collection.

was safely brought to London and formed the basis upon which was founded the Linnaean Society — a society which we will meet again with Charles Darwin.

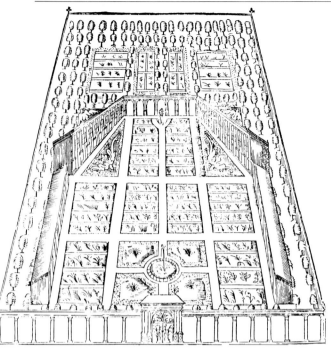

The botanical gardens at Uppsala, Sweden, which Linnaeus re-designed to separate different species, still exist today. The impression is of orderliness and permanence.

work hardened the belief in God-created, unchanging forms. The very definition of species, the great labour of classification, sought to delineate one species from the next. In principle it did not look for ancestry or evolution. It sought to divide the species, not unite them in one common origin.

Linnaeus redesigned the botanical gardens at Uppsala according to his classification plan, with the different types of plants separated into different beds. It was a visual image of what he conceived to be the Creator's plan. The ordered gardens at Uppsala, with their divided plant beds, may remind you of the postal sorting office discussed in the previous section. There the geographical divisions were seen to be 'man-made'. Similarly, in the biological world, there were those who regarded Linnaeus' divisions and classifications as more arbitrary than real. They looked at that Chain of Being, the line of organisms stretching from lifeless matter to humanity (and possibly beyond), and saw in it not the distinct groupings but the minute differences between neighbouring elements.

Le Comte de Buffon

Le Comte de Buffon (1707-88) adopted this alternative approach. He viewed species not as distinct entities but as shades on a spectrum of forms. In doing so he rivalled his contemporary Linnaeus not only in scientific theory but also in public affection.

Buffon was the keeper or director of the Royal Botanical Gardens in Paris. He was the author of one of the most massive works of natural history of all time — the celebrated *Histoire Naturelle*, which appeared in thirty-five volumes during Buffon's lifetime, with a further nine volumes (by Comte de Lacepède) following up until 1804. Buffon began the series by rehearsing again the doctrine of the Great Chain of Being:

'The first thing that emerges from this thorough examination of nature is something that is perhaps rather humbling for man; that is that he must himself be ranked among the animals... He will be surprised to see that one can descend by almost insensible degrees from the most perfect creature down to the most unformed matter, from the best-organized animal to purely brute mineral matter; he will recognize that these imperceptible gradations are nature's handiwork; and he will find these gradations not only in size and form, but also in manners of movement and generation, and in the succession of all species.'

In view of the infinite number of forms it was foolish, Buffon held, to build up classification systems such as Linnaeus had done. He argued against any arbitrary definition of species arranged to suit the classifier's whim, and was content to define a species as those animals capable of naturally interbreeding.

WAS BUFFON AN EVOLUTIONIST?

There are only occasional sections from Buffon's voluminous works relating to the issue of evolution. For Buffon, evolution was an interesting speculation and one to be entertained, but it remained largely unsupported by facts or adequate theory. He crossed swords with the Faculty of Theology at the Sorbonne over his dating of the earth, and his statements on sensitive areas are therefore studded with protective clauses!

For example, he discusses in volume four of *Histoire Naturelle* the derivation of the ass from the horse:

'Considering (the ass) . . . it appears to be no more than a degenerate horse . . . We might attribute the slight differences which exist between the two animals to a longstanding influence of climate, of food, and to the chance succession of many generations of small wild horses half-degenerate, which little by little

had degenerated still more, had then degenerated as much as is possible, and had finally produced for our contemplation a new and constant species.'

So far he appears to support evolutionary change, although the reader should note the constant use of 'degenerate'. We normally think of evolution as improving the animal form; Buffon believed that change would corrupt or distort the original type. Nevertheless, he suggests that change can occur by the influence of food and climate or simply by chance. But when he extends this notion to the whole of the animal kingdom he draws back:

'If it were true that the ass were merely a degenerate horse, there would be no limits to the power of nature, and we should be justified in supposing that, from a single being, she has been able to produce in the course of time all organized beings. But no! It is certain, from revelation, that all animals have participated equally in the grace of creation.'

In referring to the revelation of the Bible, Buffon has one eye on the professors at the Sorbonne. But is the

remainder of this quote also a sop to theological sensitivities, or is Buffon genuinely arguing against widespread evolution?

Herein lie the difficulties for the interpreter of Buffon and the reason he was dismissed by many of his contemporaries as a dilettante. Darwin never seriously acquainted himself with Buffon's work until the mid-1860s, some years after the publication of *Origin of Species*. Then it was because a colleague suggested that Darwin's views of heredity were similar to those of the Count. Darwin went away to do his homework, only to return, saying: 'I have read Buffon: whole pages are laughably like mine.'

Yet the historical sketch appended to *Origin* dismissed Buffon in a sentence:

' . . . the first author who in modern times has treated (modification of species) in a scientific spirit was Buffon. But as his opinions fluctuated greatly at different periods, and as he does not enter on the causes or means of transformation of species, I need not here enter on details.'

This point was crucial for the later Darwin. For the disciples of Linnaeus a species was an eternally distinct organic grouping; almost by definition movement across the boundaries was unthinkable and the whole classification system revealed God's original creation plan. For Buffon 'species' was just a name used to represent breeding habits, and it imposed no necessary constraints on nature itself. If, by some unknown mechanism, the natural limits in nature changed then so would our categorization of species. The classifications used by

Le Comte de Buffon, whose forty-four-volume Histoire Naturelle laid less stress on classification than Linnaeus had done.

the great systematists were not discoveries of the blueprint for creation, imposed by God from outside, but only useful tools to aid the naturalist in bringing order to his studies. Buffon expressed the aim of his *Histoire Naturelle* thus:

'Each species, each series of individuals capable of reproducing their kind and incapable of mixing with other species will be considered apart and treated separately, and we shall make no use of families, genera, orders and classes, any more than nature makes use of them.'

This breaking down of the divisions between one species and the next paved the way for evolutionary thought. But was Buffon himself an evolutionist? Much scholarly ink has been spilled arguing this question either way. He talked of changes within the animal kingdom ... and then went on to dismiss them on theological grounds. No one is certain if his dismissals were genuine or whether Buffon was being cautious at a time when ideas of evolution incurred the wrath of the authorities.

Buffon and Linnaeus represent the best-informed opinion of their time. Each set out in his own way to produce an encyclopedic account of the world of nature. Buffon's volumes had enormous lay popularity, and his departure from Linnaeus' rigid classification system opened the door to evolutionary thought. We will return to Buffon again when we discuss theories of the earth, but first we must examine a crucial relationship for the great systematists' understanding of the world: the relationship between God's revelation in the Bible and his revelation through the works of creation.

The two books

Opposite the title page of Darwin's *Origin of Species* appears the following quotation:

'To conclude, therefore, let no man ... think or maintain that a man can search too far or be too well studied in the book of God's word, or in the book of God's works; divinity or philosophy; but rather let men endeavour an endless progress or proficiency in both.'

The author is Francis Bacon, and the quotation is from his 1605 book *The Advancement of Learning*. Here is the classical statement that there are two ways of understanding the character of God, through the Bible, and through the world he has made.

Bacon went on to warn us not to 'unwisely mingle or confound these learnings together', and his intention was that each subject, the divine and the natural, should be researched separately. Yet mingled they became, so much so that the early chapters of Genesis became an authoritative guide to the geology of the earth, and the wonders of earth's creatures became living proof of the beneficence of God. The use of the Bible in science must await the next chapter, but the attempt to understand the Creator from a study of his creatures concerns us now.

John Ray was one of the first to use biology as a paeon of praise to the Creator. For the marvels of instinct in the animal kingdom, the parental solicitude of birds, the non-appearance of wasps until 'Pears, Plums, and other Fruit, designed principally for their food, begin to ripen'; for the organization of the beehive and the structure of the human body — for all these Ray praises the Maker. And such praise is not only due for the wonders of natural history but also for the beauty of the earth. The mountains are 'very Ornamental to the Earth, affording pleasant and delightful Prospects' and God has placed humanity in a 'spacious and well-furnished world'. Ray was ordained as a clergyman but he found that his conscience would not allow him to subscribe to the Act of Uniformity following the Restoration in 1660. Though debarred from practising as a priest, he still pointed men and women to God through his science.

Ray's *Wisdom of God Manifested in the Works of Creation* went through four editions in his lifetime and seventeen more by the time of

Naturalist John Ray was struck by evidence of organization in nature, of which the beehive is a supreme example. Such complex interdependence caused him to give thanks to God.

Darwin. It typified a growing flood of literature which used the wonder of nature as a window into belief in God. The high point of this approach was William Paley's famous *Natural Theology* (1802), which was subtitled 'Evidences of the Existence and Attributes of the Deity collected from the Appearances of Nature'.

Paley amassed examples from the natural world to show the intricacies of design, and thereby to reveal the wonder of the God who made them. His favourite example was the eye. The refractive powers of the lens, the turning skill of the muscles, the protective mechanisms of eyelids and tears — all these made him sure it was an instrument specially designed by God.

> 'Were there no example except that of the eye, it would be alone sufficient to support the conclusion which we draw from it, as to the necessity of an intelligent Creator.'

As we move from Ray to Paley we take a significant step. Now nature does more than point to God, nature *proves* God.

Deism

During the eighteenth century there were many who scorned the Bible as a source of knowledge. Instead they looked to mankind's own powers of reasoning to determine what we should believe and how we should act. The philosophers of this school were not atheists, disbelieving in God. But they restricted God's role simply to the opening acts of creation, denying that he subsequently intervened in human affairs. Increasingly the world was seen as a vast machine, controlled by strict scientific laws such as Newton's laws of motion.

For these 'deists' God was the designer and starter of the world machine. But once set in motion the universe governed itself. So God, they said, was no longer required and he had retreated to the wings to watch his creation. Some deists believed that God occasionally intervened but only so as to adjust the mechanism of the world. Where planets had slowly drifted off their ordained courses God interrupted to re-establish their paths.

For the more orthodox, like Archdeacon Paley, this was too great a restriction on God. It denied God's continuing providential action. The real truth, he said, was that the cosmic machine ran, not by virtue of its own inertia but by the continuing power of God. And even an occasional intervention was a poor substitute for a continuously active creator. Paley was not content with a God who stood offstage. He wanted a God who was all the time involved. What we observe as regular occurrences, the sun rising or water cascading downhill, are signs of God working behind the visible world. Without God's will, moment by moment, these things would not be. To Paley the laws of motion are our descriptions of what in reality is God's activity. Scientific laws are statements of how God works within nature.

Nature leads up to nature's God

How do we know that God exists? If there are two sources of revelation — the Bible and

Archdeacon William Paley's book 'Natural Theology' was extremely influential over many years. His argument sought to prove the existence of God from the evidence of design in nature.

nature — to which should we look for convincing evidence?

In the eighteenth century there was an emphasis on humanity's innate reason. It was better to trust in one's own reasoning than in the words of others. The thought of the eighteenth century was better, they argued, than records from the past. So, the evidence for God as deduced through studying his creation was more compelling than the uncertain voice of the Bible. Everyone could learn from nature; not everyone could trust the views of ancient books.

The case for religious belief increasingly turned away from a blind trust in the accuracy of the Bible to proofs determined from the natural world. It sometimes happens in a courtroom that the lawyers rest their case on forensic evidence and not on the doubtful memories of witnesses. And here Archdeacon Paley was the Sherlock Holmes of his day. Paley's writings are the classic expression of the argument that the evidence of order and design seen in the world point to a Designer, namely God.

In his famous book *Natural Theology* Paley uses the analogy of a watch, which points to the existence of a watchmaker (see *The*

THE ARCHDEACON FROM CARLISLE

William Paley (1743-1805) was educated at Christ's College, Cambridge — the college at which Charles Darwin was later to be an undergraduate. After graduating he worked for a time as a schoolmaster and then as a priest in various parishes in the north of England. He was a keen amateur naturalist, and his books contain many fine descriptions of organic structures and their adaptation to the environment.

Towards the end of his life he became archdeacon at Carlisle Cathedral, and three years before his death his book *Natural Theology* was published. The book is one long argument for the existence and beneficence of God, given the order and beauty of creation. The famous opening paragraphs are worth repeating in full:

'In crossing a heath, suppose I pitched my foot against a *stone*, and were asked how the stone came to be there, I might possibly answer, that, for anything I knew to the contrary, it had lain there for ever; nor would it, perhaps, be very easy to show the absurdity of this answer. But suppose I had found a *watch* upon the ground, and it should be enquired how the watch happened to be in that place,

I should hardly think of the answer which I had before given — that, for anything I knew, the watch might always have been there. Yet why should not this answer serve for the watch as well as for the stone? Why is it not as admissible in the second case as in the first? For this reason, and for no other, viz., that when we come to inspect the watch, we perceive (what we could not discover in the stone) that its several parts are framed and put together for a purpose, e.g. that they are so formed and adjusted as to produce motion, and that motion so regulated as to point out the hour of the day ... The inference, we think, is inevitable, that the watch must have had a maker.'

The argument was not new. It may be found in the writings of Plato and Aristotle and is one of Thomas Aquinas' five proofs of God in the thirteenth century. But Paley added a wealth of biological examples, and made the argument the stock response of every fellow clergyman anxious to defend his faith.

Archdeacon from Carlisle). He applies this to the natural world. Whenever we examine a plant or animal, or even the universe itself, we see such evidence of design that we are led to deduce a great Designer. And the characteristics of our natural world must highlight the nature of its Creator. So in admiring the various beauties and perfections in nature we are inevitably led to think of the beauty and perfection of God who designed them. This is natural theology, and its key concept is 'design'.

To the Victorians this was a compelling argument. While a Cambridge student, Charles Darwin had to 'get up' another of Paley's books, *Evidences*, for part of his university examination. Indeed, the book was still one of the examination options until as late as 1919. Certainly Darwin

was impressed with Paley's logic and examples. 'The careful study of these works,' he wrote in his *Autobiography*, 'was the only part of the academical course which, as I then felt, and as I still believe, was of the least use to me in the education of my mind.'

Paley imbued the study of nature with a grander purpose 'through Nature up to Nature's God,' as Pope expressed it. And the study of nature became a nineteenth-century obsession. Every Victorian lady could reel off names of ferns or fungi, every middle-class drawing room possessed a shell collection, a butterfly cabinet or an aquarium. Newspapers ran columns on natural history, natural history books were almost as popular as Dickens, and every clergyman nursed ambitions to write such works.

Victorians attend a lecture on geology. At that time there was tremendous popular interest in science and natural studies. Such knowledge was used to demonstrate the existence of a divine Creator.

When, mid-century, Darwin overthrew the careful arguments of his mentor, claiming that 'design' in nature was only a product of chance and natural causes, he was seen to be attacking not simply current biological theory but belief in God itself. If all the wonders of nature were no more than the outworkings of natural laws, was there any need to believe in a divine Designer? The watchmaker was redundant.

But Paley had another argument. Against the deists he said that you could see that God was still very much in control because he occasionally intervened in human affairs. Miracles were the hallmark of his special activity. If we find a phenomenon which cannot be explained, then we will discover that God was directly involved, transcending his normal operation of regular laws. The miracles described in the New Testament were positive proof that Christianity was true. How else could the events be explained, save the miraculous activity of God? And so he called his other famous book *View of the Evidences of Christianity.*

Here again Darwin's approach was to hit hard. He was tired of a scientific method which invoked miracles whenever a straightforward explanation seemed elusive. The subject of the origin of the earth and its species was a matter bulging with miraculous interventions. Darwin swept these all away, installing in their place the slow effects of natural law.

Mid-century, too, there arose renewed attacks on the status of the Bible. The miracles were 'explained away' and the biblical authors reduced to writing inspired prose. Some suggested the Bible was still a good book, but not The Good Book. God might speak through it, but it was not the miraculous creation it was once thought to be.

The Victorians entrusted too much of their religious certainty to arguments from miracles. And too much rested on the conclusions drawn from design in nature. The beauty of the natural world was seen as clear proof of God's existence. In responding to the critics of the Bible in the previous century they had invested almost all their theological assurance in the nature-to-nature's-God argument. They looked to the evidence of the book of God's works. When its pages were found to be loose, and the book of God's word apparently lost, there seemed nowhere to turn.

2
THE DISCOVERY
OF TIME

Archbishop Ussher, some 200 years before Darwin, calculated that the world was created by God in 4004BC. A Cambridge divine, John Lightfoot, narrowed the time down to 23 October of that year, at 9 a.m. As one commentator remarks, 'Closer than this, as a cautious scholar, the Vice-Chancellor did not venture to commit himself!'

Ussher inherited his creation date from a longstanding tradition which counted the generations from Adam to Jesus detailed in the Bible. Scholars today no longer regard these genealogies as complete, but only indicative of

A family Bible, printed in the time of Charles Darwin, has a marginal note at the beginning of Genesis referring to the creation of the world in 4004BC.

the line of descent. But in Ussher's day they added up the years to reach creation around 4000BC. To within fifty years the exact date was disputed, but 4000BC had the merit of supporting a popular idea that biblical history could be neatly divided into 1,000-year sections. In this way the six 'days' of Genesis were followed by six thousand-year periods. Psalm 90 declares, 'A thousand years in your sight are but as yesterday.' This six-times-1,000-year time-scale was interpreted by fourteenth-century Reformation leaders such as Martin Luther to mean that the world had been created 4,000 years before the birth of Jesus Christ and that humanity was now in the sixth and final 'day'.

4004 and all that

However quaint the Archbishop's views appear today, they should not be ridiculed. In the seventeenth century there was no understanding of geology and the method by which rock strata had been formed over periods of time. The Bible was, it seemed, the only source of knowledge concerning pre-history and the age of the earth. The testimony of the rocks themselves was yet to be recognized.

Yet, within a decade of Ussher's death in 1656, the naturalist John Ray stood at Bruges marvelling over a buried forest which had lain on the sea bottom and then become exposed on dry land again. This was strange, he mused, considering the recent creation of the world. If the earth was only 5,600 years old then how could there have been time for the sea to have first covered the land and then retreated?

THE DATE OF CREATION

Archbishop Ussher's date of 4004 BC became widely accepted as the date of creation. Earlier divines had suggested other dates, but their outlook was the same. The earth was of recent creation, and was now drawing closer and closer to the final judgment day. Ussher's scheme neatly partitioned history into thousand-year slots: four before the coming of Christ and two afterwards.

Whether God would allow his creation to continue to 2000 AD Martin Luther had earlier doubted. He believed society to be riddled with almost unmitigated wickedness, and so he wondered whether God would permit the rot to continue for another four centuries:

'The world will perish shortly, the last day is at the door, and I believe the world will not endure a hundred years.'

As we shall see, this view of increasing moral corruption terminated by the return of Christ was gradually replaced by ideas of progress. Initially this view still continued to count 'days' as blocks of 1,000 years: as the 'days' passed, mankind came ever closer to the kingdom of God. But this coming kingdom came to be seen as fulfilment, rather than as judgment.

Then came a crucial stage in the development of thought. As fewer people accepted God's involvement in nature, so belief in the divine ordering of history also waned. In the end the Christian framework itself was dropped, leaving mankind as the master of his own destiny.

But we must return to Ussher. What about the extra four years, to make 4004BC? This came about through Kepler's work on eclipses.

Johann Kepler, the astronomer, noting that Matthew's account of Jesus' death records that 'darkness fell over the whole land, which lasted until three in the afternoon', compared the established cycles of solar eclipses with the dating of the crucifixion. Kepler reasoned that an eclipse had caused this darkness and the nearest such solar synchronization meant the traditional dates for Jesus' birth and death had to be adjusted by four years. The birth of Christ was accordingly set back to 4BC, and it was on this basis that the Archbishop obtained his creation date.

It was a view widely accepted by the church, and printed in the margins of Bibles. Even Darwin, on being informed the date came from Ussher and not Moses, remarked, 'I low curious . . . I declare I had fancied that the date was somehow in the Bible.'

The mountain ranges also troubled him. Within recorded history the surface of the earth had changed but little. Yet:

'If the mountains were not from the beginning, either the world is a great deal older than is imagined, there being an incredible space of time required to work such changes . . . or, in the primitive times, the creation of the earth suffered far more concussions and mutations in its superficial part than afterwards.'

Ray's question was crucial.

☐ *Was the earth slowly formed over countless aeons, by gradual changes?*

☐ *Or was it shaped within more recent history by violent catastrophes?*

If the first was true, then vast periods of time were required, far in excess of the traditional 6,000-year framework. If the second, natural events must have taken place far more powerful than anything known to Ray. It was this argument — Time versus Catastrophes — that guided the geological debates of the eighteenth and nineteenth centuries. And it is necessary for us to dip a little into this geology if we are to understand the later biological theories. Charles Darwin first made a name for himself as a geologist. And he wrote scientific papers on the origins of rock formations before he tackled the origin of species.

The heroic age of geology

When Abraham Werner became a professor at the Freiberg School of Mines in 1775, the current vogue was to ascribe much of the shaping of the earth to volcanic action. In opposition to this, Werner propounded a theory

Abraham Werner, of Freiberg in Germany, built up a detailed picture of the successive layers in which earth's rocks had been laid down.

that the earth was originally covered with water, and that the rocks were formed by precipitation. A series of inundations by this briny bath had shaped the world as we now see it. Each time the waters receded, into vast caverns inside the earth, successive creations of life appeared. This accounted for the obvious geological strata, and was in agreement with the order of creation as found in the Bible. It also gave meaning to the opening verses of Genesis: 'The earth was without form and void, and darkness was upon the face of the deep.'

But Werner's thesis was not just a prop for biblical faith. It was informed by the best geology of his day. Towards the end of the eighteenth century there had been increasing interest in mineralogy, if largely for economic reasons. The Industrial Revolution had created a greater market for coal and mineral ores and the mapping and classification of rocks which resulted in a better appreciation of the earth's history. The period from 1790 to 1830 is now

viewed as the 'Heroic Age of Geology'.

To Werner the earth's crust was not a chaotic jumble of rocks, but an orderly succession of layers. Where such layers had not been disturbed the order in which they lay must be the order in which they were formed, with the oldest rocks lying at the bottom. These ideas were not new, but Werner placed them beyond dispute by his detailed and accurate field studies in the Saxony area. For those anxious to correlate the new findings with the Bible the speed of precipitation could be regarded as an open question. Those who viewed the days of creation in Genesis as marking periods of time (as did Werner himself) could adopt a slow rate of precipitation. The traditionalists, on the other hand, could still say it was all accomplished within six twenty-four-hour periods. Werner's theories dealt with the structure of the earth's crust, but could be treated as open-ended as to its age.

During his lifetime Werner's authority was unrivalled. His eloquence and charm drew students from all over Europe to his crowded classes. From here they went forth with the ardour of evangelists. In Britain the theory was championed by Richard Kirwan, president of the Royal Irish Academy, and a Swiss geologist, Jean Deluc, who had made his home in England.

Kirwan kept to a literal interpretation of Genesis — the Mosaic 'days', he insisted, were true days of twenty-four hours, and Noah's Flood was a resurgence of the original waters. From the outset he was prepared to take Scripture as a valid testimony to what had occurred at creation, and rejoiced mightily when his empirical findings corroborated the ancient record.

Deluc was more cautious. He, too, was delighted when his geological scheme agreed with traditional religious beliefs, but he claimed to have reached his views 'independent of any reference to the Book of Genesis'. It did not matter to him whether the six days referred to a literal week or to the six geological eras suggested by Werner. The testimony of the rocks echoed that of the Bible and showed belief in God to be supported by science.

Neptune or Vulcan?

Werner's geology had the merits of simplicity of structure and harmony with the Bible.

There was also a clear distinction between the immensely powerful (and, some said, sudden) forces of creation and the more recent geological forces, such as volcanoes and silting of rivers. Deluc insisted on this point. The forces of the original creation differ, he said, in magnitude and tempo to those acting in the present day. Even though the world might have been formed only slowly it could never have been by the weak and incredibly slow geological forces we can now observe. Things were different during creation. Undoubtedly God was behind both sorts of forces, those of creation and those of today, for he is both the creator and the sustainer of the universe. But there was a clear difference: the former powers were miraculous and architectural, the latter normal and incidental.

Scientifically then, the creation period and the Flood were unique, and the laws of creation were on a different scale to those of everyday

maintenance. Some geologists therefore drew back from trying to discover the earth's origins. Such moments, they said, were shrouded in divine mystery. The humble scientist could only examine the present activity of God as expressed in regular law. For details of the original creation he simply trusted the record given in the book of Genesis.

But what happened to all that water? Werner never explained satisfactorily the disappearance of the Flood, which he claimed had been sufficient to cover the highest mountains. He speculated that a passing heavenly body had attracted the water away into space. Others

NOAH'S FLOOD

In 1681, in the reign of Charles II, a prominent British clergyman wrote a theory of the earth. The title of Burnet's book offers the reader an indication of its contents:

'A Sacred Theory of the Earth: Containing an Account of Its Original Creation, and of All the General Changes Which It Hath Undergone, Or Is to Undergo, until the Consummation of All Things.'

Burnet was clear that the Bible offered the definitive statement of the earth's pre-history. The world had been created by an omnipotent and benevolent God as a theatre for the probation and redemption of mankind. The task of geology was to illuminate the main events of that creation, and in particular to catalogue the effects of Noah's Flood in shaping the terrain. The crust of the original earth had dried, Burnet claimed, opening up great cracks from which the water under the earth had poured. By the regular laws of nature the original world was destroyed and our present post-diluvian earth formed from its ruins. ('Post-diluvian' and 'ante-diluvian' mean simply 'after the

This seventeenth-century painting of Noah's Ark shows a more pastoral scene than would have been imagined in later centuries. Zoos were not yet common, and people were less acquainted with non-European species.

flood' and 'before the flood', 'diluvian' being the adjective from the noun 'deluge'.)

Burnet sought to understand the biblical Flood as the outcome of ordinary events. His starting-point was the Bible, but he understood the story of the Flood as the plan of God mediated through what we would call natural causes. The Flood had a clear place in God's providence, but from a scientific point of view it was produced by a series of natural agencies. For this he was condemned, and though in a later work he tried again to reconcile his account with Genesis, he never succeeded in throwing off the suspicion of unorthodoxy.

John Woodward (1665-1728), Professor of Medicine at Gresham College, London, agreed that the Deluge had revolutionized the face of the earth. Yet such a cataclysm could scarcely have been produced by the ordinary operations of nature. It demanded a divine decree. Wind, water and volcanic action could not, by themselves, substantially alter the face of the earth. The Flood was a miraculous intervention by the Almighty.

During the eighteenth century the debate continued with different writers emphasizing different points. To some the Flood was the only sufficient agency, to others there was a place for volcanic action and the constant wear of tide and rain. To some the shape of the terrain was a product of purely natural causes, to others the earth was a stage for divine interventions.

returned to an eighteenth-century belief that it had vanished once more into the interior of the earth.

The 'Neptunist' theory, as it was called after the Roman god of the sea, dominated the first decade of the nineteenth century. But as geological studies multiplied it became increasingly difficult to maintain. Faced with increasingly complex evidence, belief in the efficacy of a single worldwide deluge was replaced by schemes involving a whole series of such cataclysms. It was like a murder investigation: as more and more footprints were discovered in the earth, so the evidence demanded a more complicated view of what had happened on the fateful night.

As regards creation, Alcide d'Orbigny counted twenty-nine geological catastrophes, and only the last of these was Noah's Flood. Even then there were difficulties and an alternative, more simple, thesis began to win acceptance. A rival school, named 'Vulcanists' after the god of fire, stressed the effect of heat rather than water, of volcanoes rather than floods, in forming the earth. Because they could not slip in Noah's Flood as the last of the catastrophes, they could not claim the support of Scripture. Yet their views on the gradual formative powers of the earth's heat made sense.

At times the debate was acrimonious. In Edinburgh a play was hissed off the stage when its Vulcanist philosophy offended a house packed with Neptunists! But though the geological agencies involved altered from one theory to the next, a new perspective lay at the back of them all: Time. Whether the rocks were seen as 'sedimentary' (laid down by water) or 'igneous' (laid down by processes of heat), they obviously took a long time to form.

Gradually, like the rock deposits themselves, it came to be accepted that the creation of the

Volcanic activity is an equally evident contributor to the shape and distribution of geological features. But, asked early-nineteenth-century geologists, which was the more important, water or volcanoes?

There has never been any doubt that water has made a considerable impact on geology. One obvious example is coastal erosion.

world had taken thousands of years. They had discovered Time.

The fossils come alive

The presence of marine fossils on dry land, even on the tops of mountains, caused no embarrassment to the catastrophists. As we have already seen, fossils were once regarded as mere 'sports' of nature, unrelated to the living organisms they resembled. In flood geology, they were seen as the ossified remains of organisms left as the waters receded. But the fossils also witnessed to a great passage of time and palaeontologists (students of fossils) were adding their own evidence on the antiquity of the earth.

In France the leading scientist Georges Cuvier (1769-1832) declared certain animals had indeed become extinct. Until the turn of the century people believed that unearthed bones relating to strange, and sometimes enormous, animals belonged either to giants living before the Flood or to species which had long since migrated to other parts of the world. From their bones Cuvier was able to reconstruct models of these long-lost animals, and continuing geographical exploration suggested that such creatures had mysteriously vanished from the earth.

The structure of the fossil layers in the Paris Basin convinced Cuvier that the deposits had gradually built up over a long period, though how long he was unable to say. Some of the valleys suggested they had been scoured out by running water, but the volume of water required was formidable. Cuvier drew back from the abyss of time which seemed to open before him:

> 'In geology we must limit ourselves to the observations of facts, since the hypothesis which appears the simplest and most natural is subject to insoluble difficulties for the present.'

Where Cuvier was unwilling to speculate, perhaps out of religious sensitivity, Georges

The study of fossils, such as this dinosaur in a cave in America, has long been an important key to knowledge of the distant past.

Louis Leclerc, Comte de Buffon, was prepared to calculate. The solar system, he believed, had been formed out of matter ejected by the sun, following a collision between the sun and a comet. In the beginning the earth, like the other planets, was a molten sphere. By observing the rate of cooling of an iron globe, Buffon claimed it had taken almost 75,000 years for the earth to attain its present temperature. This view, when published, did little to cool the tempers of the theologians! Buffon was forced to preface the fourth volume of his *Histoire Naturelle* with a formal retraction of the heretical views expressed in the first volume:

> 'I have no intention of contradicting the text of Scripture . . . I believe firmly everything related there concerning the creation . . . and I abandon whatever concerns the formation of the earth in my book, and in general everything which could be contrary to the narration of Moses, having presented my hypothesis concerning the formation of the planets only as a pure supposition of philosophy.'

Supposition or not, the belief in an age for the earth of thousands of years was gaining credence. In Britain the growth of canal building made information on rock strata vital. William Smith, known popularly as 'Strata' Smith, was a mining engineer who made extensive studies of the fossil remains revealed as canal cuttings were fashioned. A self-made man, Smith never regarded himself as a scientist, and neither belonged to the Geological Society of London nor cared for the controversies then raging among its members.

He gathered information to further his own business. But in so doing he outstripped the academics in field work and his *Map of the Strata of England and Wales* (1815) made it clear that the strata were laid one on top of the other like so many slices of bread. London clay (the youngest) was at the top and granite was at the bottom.

Furthermore, each layer contained characteristic fossils. To descend underground was to pass back along time, with the succession of life forms as the key to the ages. The time-scale required for the formation of the rocks and their embedded fossils exceeded that allowable in a literal interpretation of Genesis. The rocks spoke of thousands of years, not six days.

Buckland and the Dean

It is often said that Charles Darwin destroyed the Victorians' belief in the Bible. However, the debates on the place of the Bible in a scientific world took place in the decades before Darwin's work. Since God had revealed himself both in nature and in the Bible it seemed inconceivable that the two should disagree. Genesis and geology were one.

But the new theories upset the traditional six-day time-scale of Genesis. Not since Kirwan's day had professional geologists tried to defend a literal reading of the first chapter of Genesis. All assumed vast tracts of time in the formation of the rocks. How, then, could Christian geologists, a number of whom were clergy, square their science with the Bible?

Prominent among their number was William Buckland, Canon of Christ Church Cathedral, Oxford, and later Dean of Westminster. In 1819 he became professor of the new chair in geology at the university. He was a distinguished teacher, and his lecture room, a terrible jumble of rocks, skulls and skeletons, was famous all over the university. His habit of conducting field studies, whether in caverns or on hilltops, dressed in top hat and academic gown, added a touch of eccentricity. At home he entertained his guests with such dainties as horse's tongue, alligator and ostrich. One guest complained, 'Party at the Deanery; tripe for dinner; don't like crocodile for breakfast!'

However idiosyncratic his menu, his geology was widely regarded as definitive. He admitted that the creation had taken many thousands of years, but he also fervently believed the fossil beds revealed a final Flood which had killed countless animals and embedded their remains in the sediments. Important, too, was the lack

During the early nineteenth century canals were excavated throughout England. This gave ample opportunity to a man such as William 'Strata' Smith to study rock formations.

of human fossils in the more ancient strata. Mankind was of recent origin, just as Moses recorded in Genesis. Noah's Flood had actually happened — so said the Bible, and so said the sediments and buried fossils.

Nevertheless, for all his attempts to marry his science with the Bible, he was criticized for abandoning the literal six days. The 1844 meeting of the British Association in York was the occasion of an outspoken attack by the local Dean on the errors of Buckland and other leading geologists. One of those pilloried remembers:

> 'The Dean (of York) attempted to explain the Mosaic cosmogony literally. Marine volcanoes, he thought, together with the supernatural rain of the Flood, had deposited all the strata, as we see them now, in the course of a few days; and the embedded fossils represent the remains of animals that were so obliging as to die in the definite and regular order in which their shells and bones are now deposited.'

But the views of Dean Cockburn were, by then, in the minority, at least among practising geologists. Even by the 1830s there was widespread acceptance among geologists

MATCHING GEOLOGY WITH THE BIBLE TODAY

There are many Christian geologists who still try to harmonize the message of revelation with the testimony of the rocks. The same nineteenth-century harmonies are produced. Some writers change the 'days' of Genesis to geological and evolutionary epochs, or interpret Genesis as a vision. Others adopt the theory that the first verse of the Bible speaks of a long but indefinite time, during which much of the world's geological structure

was formed — only after this long period was there organic creation, which occupied six days, and only these six days does the Bible record in any detail.

All these viewpoints are still held. Fundamental to them all is the conviction that, since all truth is one, the message of both science and the Bible must be in straightforward agreement. Similarly, proponents of these views refuse to be bound by a principle which appears to them both materialistic and arbitrary, that of belief in an unbroken web of cause and effect. Why should there not be special forces at work in creation?

Whether their harmonizing attempts get near the truth is a matter for Bible exegesis and scientific analysis.

It is not a question which can be answered here, though there is more about it in the final chapter. But the principles behind the opposing viewpoints need to be grasped, for they underlie more recent debates between six-day creationists and evolutionists. The nineteenth-century geological debates might have begun with arguments over the interpretation of Genesis, but the underlying questions on belief and the nature of science were more important.

☐ *In what sort of God do we believe? Magician or Grand Designer?*

☐ *In what sort of science do we believe? A science bound by the limits of cause and effect, or one flexible enough to permit the intervention of God in miracle?*

that the earth was very old. Yet Dean Cockburn was raising a fundamental question:

☐ *Should the geologist start with a literal reading of Genesis and interpret the rocks accordingly?*

☐ *Or should he use his knowledge of strata to illumine a more poetic understanding of the Bible's creation story?*

If God was the author of both Word and World, which should take preference over the other? Which should be the starting-point?

Even the devout could not deny that the rocks indicated the passage of innumerable years rather than a single week. Attempts were made, as we have seen, to harmonize the scriptural and geological accounts by suggesting the 'days' of Genesis represented geological ages. Others claimed the Bible only mentioned God's final act of creation: the ages in which the rocks were formed were inherent in the opening verse of Genesis, 'In the beginning . . . ' This was Buckland's argument. The word 'beginning' expressed 'an undefined period of time' during which there had been a long series of operations to form the earth's structure. Since these were irrelevant for the

biblical story they were simply 'passed over in silence by the sacred historian'. Between verses one and two of the first chapter of Genesis lay innumerable years.

Hugh Miller, one of the most popular geological writers whose *Testimony of the Rocks* had sold 42,000 copies by the end of the century, adopted a less literal approach. The Genesis days, he said, were days in which God revealed to Moses, in a vision, his creation of the world. As such the timing did not relate to the actual formative processes God used, but to the days of revelation to his servant Moses.

In such ways Christians sought to come to terms with geology. Whatever their particular interpretation, they saw harmony between their scientific views and the Bible. There were divergent views on interpretation, as there were over particular geological theories. But it was not a battle between science and religion, for there were proponents of both on each and every side!

The importance of the religious question raised by science was strongest in England. In France the Revolution had established secular institutions, and men such as Cuvier no longer feared the ecclesiastical authorities as Buffon had done. But in England you needed to be a clergyman to be a fellow at either Oxford or Cambridge University, and satisfactory posts as professional scientists were still rare.

Therefore the men of science in England were often clergymen. At Cambridge the Rev. J.S. Henslow was professor of botany, the Rev. William Whewell was professor of mineralogy, and the Rev. Adam Sedgwick was Woodwardian professor of geology. Just before he sailed for South America, the young Darwin went on a geological field trip with Professor Sedgwick. It was one of the best courses he could have received for Sedgwick was acknowledged as the country's leading geologist. At the rival university Oxford, there was William Buckland. And at Oriel College the professor of geometry, the Rev. Baden Powell, also wrote on scientific matters. Now all these men had differing views, both in science and religion, but they were all convinced of the importance of truth in both disciplines. They sought to accommodate their science and their Christian faith in different ways, but there was never a battle over whether or not God was the creator and upholder of the natural world. 'Billy Whistle' (as Whewell was known to his undergraduates) was less conservative than 'Robin Goodfellow' (Sedgwick), but he was no less a believer.

The geological debates never questioned the existence of God. But, as Darwin was to realize, the new geology had admitted one fundamental concept: Time.

Speaking of the time needed for the slow process of evolution to work, Darwin was later to write in *Origin of Species* of the work of a geologist called Charles Lyell:

'It may be objected that time will not have sufficed for so great an amount of organic change, all changes having been effected very slowly through natural selection. It is hardly possible for me even to recall to the reader, who may not be a practical geologist, the facts leading the mind feebly to comprehend the lapse of time. He who can read Sir Charles Lyell's grand work on the Principles of Geology, which the future historian will recognise as having produced a revolution in natural science, yet does not admit how incomprehensibly vast have been the past periods of time, may at once close this volume.'

Without long periods of time any theory of animal descent and gradual modification was doomed. Darwin needed time, millions of years in place of Ussher's 4004 BC. It was the geologists who gave it to him.

Lyell and the uniformitarians

By the 1830s the geologists had bequeathed an understanding of time and had questioned a literal interpretation of Genesis. It may seem that this was a sufficient gain from geology, and

Charles Lyell, author of 'Principles of Geology', had a central influence on Darwin's thinking. He helped him to think of 'natural causes' rather than frequent divine interventions.

that we should now return to the main story of biological evolution.

But for Darwin this was not enough. Central to his thesis would be a claim that evolution occurred by purely regular causes, by the cumulative effect of innumerable small steps. In *Origin* there was to be no place for overpowering divine interventions. But the best science of geological origins was full of 'catastrophes' and the insistence that the forces of creation differed from those of the present. A change in scientific thinking was called for.

Again it was the geologists who came to his rescue. Through Charles Lyell they provided the scientific understanding and methodology that made possible the later theory of evolution by natural selection. We must see what they said.

Lyell (1797-1875) was the eldest son of a rich landowner and was educated at Oxford. Originally he intended to enter law, but at the age of twenty-one his interest in geology was suddenly awakened during a trip with his parents through Europe. Within the year he was a member of the Geological Society. He met Cuvier in Paris and, more importantly, a lesser-known professor of mineralogy and geology called Constant Prévost. Prévost had criticized Cuvier's emphasis on catastrophes, preferring to believe that geological changes in earlier ages did not differ essentially from the events currently taking place on the earth. Prévost was a shy man and did not want to wrangle with Cuvier, but Lyell had no such timidity and at once set about assembling the evidence for gradual changes in the earth's structure. By 1830 Lyell had completed the first part of his *Principles of Geology* in which he expounded the theory of uniformitarianism.

Lyell's theory was not new. It had been proposed forty years earlier by a contemporary (and opponent) of Abraham Werner, a Scottish physician called James Hutton. But Lyell used the considerable field data garnered since Hutton's day to support the theory with evidence, and so rescue it from obscurity. In time it became the dominant geological view, and Lyell is often known as the founder of modern geology.

The uniformitarians dismissed the need for 'catastrophes'. The seventeenth-century belief that Noah's Flood was responsible for everything. This theory had now multiplied into whole sequences of cataclysms. As evidence piled up, the story of how one catastrophe had been followed by another became ever more complex. The uniformitarians simply believed it to be a lie, and wanted to reduce the whole lot to a few straightforward laws.

This shift in viewpoint echoes the seventeenth-century dispute over whether or not the earth went round the sun, or vice versa. The traditional view of the Middle Ages placed the earth at the centre of the universe. But as astronomers began to plot the course of the stars they were forced to apply special correction after special correction in order to express all the movements as orbits around the earth. And the special mathematical constructions they employed differed from planet to planet and from one otherwise unaccountable phenomenon to the next. Copernicus could not fault their astronomical predictions (the scheme, complicated as it was, worked) but he was upset by their mathematics. He wanted one central principle, namely uniform motion in a circle. Aristotle had taught that such movement was the most perfect, and this befitted the creation. So Copernicus introduced a system in which the sun was at the centre of the universe and all the planets revolved around it. Since he was committed to circular motion (but we now know the planets move in ellipses) the final scheme was even more complicated than what had gone before. But at least his principle of uniform, regular motion was upheld. And this solution attracted Copernicus not because he had looked through a telescope, but because it was mathematically more elegant.

In a similar way the uniformitarians in geology brought in a new focal principle. They too wanted regular laws which controlled everything, rather than specific solutions for each unique occasion. The early earth had been formed, they said, by the incessant action of geological forces we still see operating today. Winds wear away mountains, glaciers transport huge boulders, volcanoes erupt and wreak topographical change. Valleys are eroded by rivers, cliffs by the sea; and the material washed away is deposited to build up a different land elsewhere. These natural forces, and these alone, were sufficient to explain the formation of all geological features. Admittedly vast amounts of time would be needed, but once this was granted any need for world-shaking cataclysms or miraculous and divine interventions was removed.

This brought the uniformitarians into disagreement, of course, with the catastrophists, for whom it was axiomatic that present-day forces were simply inadequate to explain all geological phenomena. A greater power must have been at work, said the catastrophists, the direct power of God. Not so, replied Hutton and Lyell. God may have worked in the past as he works today — through the agency of normal geological effects. Given time the accumulative effect of small forces is equal to that of a sudden 'catastrophe'. We all know the damage that can be caused by a dripping tap.

Lyell's approach did not guarantee success at every stage of his argument. As with Copernicus, the data sometimes had to be forced to fit the central principle. And in his attempt to establish the principle he overstated his case. He insisted that natural forces have always been of the same nature and *intensity*. Most geologists did not clash with him over his emphasis on natural causes. They tended to reserve the world-forming catastrophes for the moments (or some said epochs) of creation. And even then geologists like Sedgwick looked more to natural events, however unique, rather than to direct miracles. So they welcomed Lyell's demonstration of the power of gradual cumulative forces.

But his opponents could not join Lyell in the further step of believing that these slow and gradual forces were the only ones available to the Creator. They failed to see why past forces could not have differed markedly from ones we now observe. For example, the Mezozoic rocks of southern England contain bands full of marine fossils, and to the early geologists these fossil 'graveyards' screamed out for the

Rocks are shaped, over long periods of time, by several different means: volcanoes, flowing water, ice, sometimes simply the wind.

Flood. Much as today some scientists attempt to explain the extinction of the dinosaurs 65 million years ago by theories of passing comet showers or meteors, so catastrophists sought to explain their data with similar worldwide cataclysms. When, later, theories of an Ice Age in Scotland were propounded by Louis Aggasiz, they were dismissed by Lyell. His geology could not accommodate such severe climatic changes. Aggasiz was a follower of Cuvier and his ideas smacked of catastrophism. Lyell's commitment to a rigorous uniformity blinded him to ideas we now believe to be correct. In the edition of his *Principles* published after Agassiz's visit to Britain he only gave a passing reference to glaciers, and it was not until many years later that he was finally convinced.

A charge of atheism

Both catastrophists and uniformitarians agreed that the world had taken many thousands of years to form. They differed over the processes of formation, but neither was driven by the need to square their views with a literal understanding of Genesis.

And yet Lyell and his followers were accused of atheism, because in the eyes of some critics they denied the intervening hand of God. If there were no catastrophes, no miraculous interventions, where was God? And if God was not involved in the running of nature, could we be certain of his concern for the affairs of humanity? Were not God's interventions in the ordering of the world proof that he cared for his creation?

The 'irreligious' uniformitarians also failed to emphasize the lessons of divine design. We saw in the last chapter how the Victorians loved to highlight the intricacies of nature as evidence of God's plan. In geology Buckland praised the wisdom and goodness of the Omnipotent Architect who, by a happy distribution of coal and iron, had assured manufacturing primacy to his British creation! However ludicrous (and even arrogant) this belief, behind it lay an understanding of science as the revealer of God's bounty. In contrast, the new science only sought to give a description of the forces at work in nature, with no deductions as to the plan or intentions of God. These scientists neither looked to God for miracles nor peppered their writings with exclamations at the wonder and usefulness of creation. Slowly they began to sever the link, previously explicit, between Creator and creation.

Yet, for many, uniformitarianism exalted God rather than denied him. The stress in the

The gardens at the Palace of Versailles in France give vivid physical expression to the eighteenth century's passion for order and regularity. This outlook inclined people to believe in a God who worked through consistent laws more than through miracles.

eighteenth century had been on a God of reason and regularity. The motions of the planets reflected the nature of God as the creator of a perfect machine. For God to intervene through miracle was like a mechanic tinkering with a faulty part: it showed weakness in the original design. One eighteenth-century scientist wrote:

> 'We think him a better Artist that makes a Clock that strikes regularly at every hour from the Springs and Wheels which he puts in the work, than he that hath so made his Clock that he must put his finger to it every hour to make it strike.'

A twentieth-century analogy would be the flying of a 747 Jumbo jet. For most of its flight it is controlled by an automatic pilot, programmed according to a pre-determined flight plan. Only at critical moments, at landing and take-off, does the pilot become directly involved. The advance of technology to include automation even of these critical moments is a testimony to human skill and inventiveness. Perhaps the eventual aim is a pilot-less plane. Yet to take away the captain from the flight deck would make many of us feel less safe, not more! Where one person needs to know that the pilot is at the controls, another marvels at the cleverness of the engineers that made auto-piloting possible.

So, in the nineteenth century, there were scientists who wanted to extend God's 'automatic piloting' of the world to include even the critical moment of creation. But this left the feeling that God was no longer in immediate control. Had he left the flight deck?

Buckland wanted reassuringly to point to signs of God's direct involvement. Geology, he wrote, provided 'direct and palpable refutation'. God had repeatedly intervened in nature, 'not

FOSTERING INDOLENCE

Many early geologists regarded the creation of the world as a subject almost beyond scientific investigation, let alone explanation. If God had created through a series of miracles, then it was sufficient to say, 'God made it so'. Science simply catalogued the impact of this divine power.

Opposed to them were those who attempted to explain all phenomena in terms of natural laws and causes. But to understand the past they had to assume that the natural causes active those many years ago were the same as they observed in the present. This is well expressed in the sub-title to Lyell's *Principles of Geology* (1830-33):

'Being an Attempt to Explain the Former Changes of the Earth's Surface, by Reference to Causes Now in Operation.'

They argued by analogy from the present to the past, and elevated to an axiom the principle that the natural forces at work in creation operated in the same manner, and with the same intensity, as today. Here there was no place for catastrophe or miracle, only the unvarying web of cause and effect.

Of course, to reach back beyond even the first creative forces to a time when there was no earth at all was to exceed the scientist's brief. Hutton ended his book *Theory of the Earth* (1795) by concluding, 'We find no vestige of a beginning — no prospect of an end'. He did not so much deny the biblical creation account in Genesis as ignore it. The earth, Hutton and the uniformitarians claimed, yielded no evidence of its ultimate origin, only the everyday marks of constant wear and change.

Seas ebbed and flowed, deposits were laid down only to be eroded in later years.

This uniformitarian stance was criticized for adopting a principle that could not be proved. How do we know the past was like the present? Why should God not use violent catastrophes, either of natural or supernatural origin, to shape the earth? Lyell was prepared to acknowledge these objections, but he believed that the all-too-ready recourse to miracles by the catastrophists was a stumbling-block to scientific advance. To treat the study of origins as a mystery, or to simply declare that 'God made it so', was a hindrance to understanding the possible causes behind phenomena. If Buckland was fearful that without cataclysms there was no sign of God, Lyell was equally worried that without the principle of uniformity there would be no science.

He concluded Volume 1 of his *Principles*:

'Never was there a dogma more calculated to foster indolence, and to blunt the keen edge of curiosity, than this assumption of the discordance between the ancient and existing causes of change . . . The student, instead of being encouraged with the hope of interpreting the enigmas presented to him in the earth's structure, was taught to despond from the first. Geology, it was affirmed, could never rise to the rank of an exact science; the greater number of phenomena must forever remain inexplicable . . . '

All he asked, as he explained in a letter to a colleague, was 'that at any given period of the past, don't stop inquiry when puzzled by refuge to a "beginning".'

Today, especially in schools in the United States, there are attempts to teach science based on a near-literal reading of Genesis. For many the objection to this is not religious, but scientific and educational. Is it good

science unquestioningly to invoke the miraculous when no known natural agent seems possible? Should we not rather admit our ignorance and pursue our studies in the hope that one day the jigsaw of natural causes will be complete? This was Lyell's urging, and today many scientists are unwilling to be tied by Genesis or resort to miracles when natural explanations are elusive.

Thomas Huxley, known to us as Darwin's Bulldog, recalled his scientific pursuits in the first half of the nineteenth century:

'I had set out on a journey, with no other purpose than that of exploring a certain province of natural knowledge; I strayed no hair's breadth from the course which it was my right and my duty to pursue; and yet I found that, whatever route I took, before long, I came to a tall and formidable-looking fence. Confident as I might be in the existence of an ancient and indefeasible right of way, before me stood the thorny barrier with its comminatory notice-board —

"No Thoroughfare. By order. Moses."'

Those who invoked miracles accused the uniformitarians of adopting an unproved principle. Why should they discount miracles as part of their scientific explanations? But the uniformitarians in turn retorted that at least such a principle allowed them to investigate God's world.

At stake was a new viewpoint on science. The old science invoked divine will and miraculous cause as an explanation of the unknown. The new science postulated yet-to-be-discovered laws.

The one hindered research because such mysteries were unlikely ever to be clarified. The other held open the hope that, one day, they would be. It was dissatisfied with what were regarded as inadequate explanations.

blindly and at random, but with direction to beneficial ends'. Therefore, 'we see at once the proofs of an overruling Intelligence'.

But Hutton and Lyell could see none of this. Hutton himself by no means denied God's original creation, nor his subsequent running of the universe through natural law. He recognized that the 'cosmic programme' was of divine origin and that so-called natural agencies were, like flight instructions, reflections of the will of the creator. The world spins on its axis because God wills it; the seas ebb and flow, the mountains rise and fall, for the same reason. Though undoubtedly his and Lyell's views could be used to exclude God from his universe, for others they expressed a grander concept of God than the interfering magician proposed by Buckland. The regularities we observe in nature are but human summaries of God's regular activity. As Thomas Chalmers expressed it, 'The uniformity of nature is but another name for the faithfulness of God.'

In 1832 Lyell was appointed professor of geology at the newly established King's College in London. The college was formed with a strong religious bias in direct opposition to neighbouring University College in Gower

Thomas Chalmers was a leading Scottish churchman of the mid-nineteenth century. He pioneered social and ecclesiastical reform, all as an expression of his belief in the constancy and dependability of God.

Street. The 'Godless of Gower Street', as they were called, were taught the new doctrines of Jeremy Bentham, the philosopher of utilitarianism. His skeleton, dressed in his own clothes and positioned in a glass box, still today adorns one of the hallways. But King's had a church tradition, and the governing board was initially hesitant about appointing Lyell. However, by the end of his second lecture they were content. As Lyell records,

> 'I worked hard upon the subject of the connection of geology and natural theology, and pointed out that the system which does not find traces of a beginning, like the physical astronomer, whose finest telescope only discovers myriads of other worlds, is the most sublime.'

He found his reflections of divinity in the system of nature, rather than in its creation. He ended by quoting the Bishop of London to the effect that truth always added to our admiration of the Creator.

Perhaps Lyell, like Buffon before him, had an eye on the authorities. But this was not really a battle between science and religion. It was a debate within religion on the nature of God and within science on the extent of natural law. There was almost universal agreement that the world was created by God.

A secular world

The Latin for 'time' is *saeculum*, from which we derive our own word 'secular'. A 'secular' world-view is one that limits explanation and horizons to this world, and this time, only. God, who inhabits eternity rather than time, is not acknowledged. In the search for geological time, measured in years, the scientists began to restrict their questioning to this world's 'time', agreeing to be bound by a secular understanding of cause and effect.

Traditionally God had been understood as the First Cause behind all phenomena. He was their author. Natural forces were viewed as his instruments and called 'secondary causes'. But

The clothed skeleton of philosopher Jeremy Bentham sits in a case in University College, London. Bentham was a key figure in founding this college, renowned as one of the first educational institutions not to have a specifically religious basis.

during the nineteenth century, a gradual shift took place in the way earth's origins were explained. Now they were expressed solely in terms of these latter, secondary causes. Any mention of God's design or immediate involvement was tacit. From there it needed just a short step before people stopped talking of 'secondary' causes and spoke only of 'natural' causes. If believers understood that behind scientific laws and mechanism was the hand of God, then that was their personal commitment. It was left to the eye of faith, not the research of the scientist. God's creative Word was replaced by a secular understanding of time.

Now, over 100 years after Darwin, we no longer talk of primary and secondary causes, only of 'natural' causes. And this can be misleading. In the popular mind a natural cause implies something which works by itself and not by the hand of God. The very term rules out God. But when Darwin talked of the laws of science or nature he was not committing himself as to their ultimate origin. Nature may run according to self-sufficient causes (what we now call 'natural' causes) or to causes with their origin in the will of God. Darwin had his own personal views, but his published scientific work only described the manner in which nature worked and left the deeper questions untouched. In the remainder of this book we will need to talk of natural causes, for that is the phrase used today. But we, like Darwin, do not mean to imply that these are godless causes, only that they are explanations within nature.

As we have seen, geology raised questions about the interpretation of Genesis. Were the six days literal periods of twenty-four hours, do they represent geological epochs, or are they connected with the revelation given to Moses? However, the catastrophist-uniformitarian debates were asking a deeper question — in what sort of God do we believe?

☐ *Do we believe in a God who intervenes in the course of nature — a miracle-working God?*

☐ *Or do we place our trust in a God who works solely through the agency of natural law?*

Are we content to believe in a God who stays unseen behind the so-called natural laws, a God acting solely through law and not in intervening miracle?

For some people, a God of regularity and order, a God committed to working through 'natural' agencies, is more attractive than a God who constantly intervenes. Others, perhaps less confident, need to see a God of miracles if they are to believe. At root the nineteenth-century question was 'How did God create the world — by miracle or mechanism?' The answer given depended only partially on the scientific data. It also reflected the prior attitudes and beliefs of those weighing the evidence.

3
THE FORERUNNERS
OF DARWIN

To describe Erasmus Darwin, grandfather of the more famous Charles, as a lunatic is accurate, but misleading. Born exactly 100 years before his grandson began his epoch-making voyage, Erasmus remains the most genial, ebullient and idiosyncratic of the Darwin family. His lunacy relates not to his state of mind but to his membership of an eighteenth-century learned society, the 'Lunar Society', founded by Matthew Boulton around 1766. The group gathered at the house of one or other of its members once a month on the night of the full moon, so that the members could the more readily find their way home. The society existed to promote the arts and sciences, and although its membership list never rose much above a dozen names it included many now famous in the history of science. Watt, the developer of steam engines, Priestley the chemist, Murdoch, the inventor of gas lighting — all these joined with Darwin, and were nicknamed by him 'lunatics'.

Erasmus was a successful doctor, first in Lichfield and later in Derby. His reputation was such that George III suggested he might come to London as a Royal Physician — an unofficial invitation which, to the delight of Derby, he turned down. His interests included animal and plant biology, technology and poetry, and when his medical reputation was firmly established it was to these he turned. He published first a long poem entitled *The Botanic Garden* which praised 'the immortal works of the celebrated Swedish Naturalist Linnaeus'. It was slated as a pompous rhyme by Lord Byron, but its popularity saw it through numerous editions.

In 1794 and 1796 his two-volume *Zoonomia* or 'The Laws of Organic Life' came off the presses. It was translated into German, French, Italian and Russian — and placed on the Papal Index. At the turn of the century it was an important talking-point, and introduced the word 'Darwinism' long before Charles was even born. Erasmus' intention was to write

'Darwinism', at the beginning of the nineteenth century, referred to the views not of Charles Darwin but of his grandfather Erasmus, whose 'Zoonomia' put forward evolutionary views.

a medical textbook to 'unravel the theory of diseases', but the book is remembered now only for its explicit evolutionary views. He spoke of animals undergoing 'perpetual transformations'. Regrettably the style was more speculative than scientific, and the book would have become one of the minor curiosities of scientific history had not his grandson formulated a much more comprehensive theory. Coleridge coined the word 'darwinizing' as a contemptuous reference to Erasmus' habit of indulging in wild speculation. In his autobiography Charles Darwin expresses his disappointment on reading his grandfather's work, the 'proportion of speculation being so large to the facts given'.

Speculation versus facts

'Speculation against facts' may be taken as a theme underlying the views on evolution in the late eighteenth and early nineteenth century. Everywhere there was speculation, for notions of 'development' were to be found everywhere. This is a book on biological evolution so that will remain the main theme, but in the nineteenth century the greater revolution lay in a new appreciation of change and history. Evolution was simply history of the natural world, and only a sideshow in the wider development of historical consciousness.

Ultimately, the sideshow was to become the main attraction as philosophers and politicians used the new biological theories to undergird their general beliefs in historical change. We will consider all this in a later chapter, but at the start of the nineteenth century ordinary men and women were beginning to appreciate history for the first time.

Until the close of the previous century the European mind was dominated by the classical view of history. Age might follow age, but if there was an improvement in people's lot from one year to the next it was only a temporary Golden Age. In due course history would revert and, as though tracing a giant circle, humanity would be back where it was. No new order of things had ever arisen or would arise; there was only a recurrent cycle of decay and of restoration. As the Roman emperor Marcus Aurelius expressed it, a man of forty years has seen all that has ever been or shall ever be. The Old Testament book of Ecclesiastes breathes the same air when the author writes:

> 'What has happened will happen again, and what has been done will be done again, and there is nothing new under the sun. Is there anything of which one can say, "Look, this is new"? No, it has already existed, long ago before our time.'

Books of 'history' there were, but they were attempts to quarry from the past moral or philosophical truths relevant for the day. There was little attempt to understand or empathize with the past, no reconstruction of Roman villas or Viking villages 'just to see how they lived'.

Edward Gibbon (1737-94) wrote his famous six-volume *History of the Decline and Fall of the Roman Empire* between 1776 and 1787. The first volume had a magnificent reception and soon found its way, Gibbon informs us, to 'every table' and to 'almost every toilette'. Gibbon based his books on a study of the original materials and the writings of European historians. It was the start of a trickle of historical works which, by the mid-nineteenth century, had turned into a flood. The early nineteenth-century public became history-conscious, and in his *Waverley* series Sir Walter Scott developed the historical novel to satisfy the new thirst.

Museums were founded and stocked, artists painted historical scenes on vast canvases, architects scorned the traditional classical buildings and rediscovered the gothic style of the Middle Ages. A historical approach was also applied to the Bible and, as we shall see later, this was to play a major part in the church's antagonism towards Darwinism.

In the late eighteenth and early nineteenth century Britain was also experiencing the Industrial Revolution. Like two wheels on the same axle the revolution in historical understanding was coupled with that in industry and science. The new steam age took people

EARLY HINTS ON HEREDITY

Robert Chambers, the anonymous 'Mr Vestiges', was fascinated by ideas of evolution and, as we will see, he was himself a 'freak'. Interestingly, another six-fingered family had already played a part in the search for biological knowledge. A century before Chambers, the Berlin surgeon, Jacob Ruhe, had been born with an extra digit on each hand and foot. Though his father had been perfectly normal, Ruhe shared this peculiarity with his mother and his maternal grandmother. Three of his seven brothers were affected, as were two of his own children. The Director of the Academy of Sciences, Pierre Louis Moreau de Maupertuis, was introduced to the surgeon and immediately recognized the inevitability of some form of hereditary transmission within the Ruhe family.

In the mid-eighteenth century, knowledge of the mechanism of heredity was hazy, to say the least. Intrigued by his hexadactyl friends, Maupertuis embarked on a series of animal-breeding experiments. Here he stumbled on one of the concepts geneticists use today, namely that each parent contributes a set of 'particles' to form his or her progeny. Within this scheme, certain particles can be more dominant whereas the effects of others only reappear after several generations.

Sadly, Maupertuis fell out with Voltaire, whose satirical pen soon demolished the surgeon's reputation far more effectively than his advocacy of strange, and still unacceptable, theories. Maupertuis' work was known to Buffon, but in general it lay neglected save that it reinforced the idea that inheritance somehow involved the characteristics of both parents.

Charles Darwin initially held that offspring were 'blends' of the two parental natures, much like mixing two pots of paint together. This understanding placed him in a difficulty with regard to changes. How could changes occur, and more particularly how could they be propagated? In terms of blending, a change in one generation will slowly be removed as the generations pass, just as mixing magnolia paint with countless 'generations' of normal white paint will gradually remove the magnolia colouring. Had Darwin used Maupertuis' idea of particles remaining over many generations he would have been less troubled by the problems of inheritance.

---

from the villages into growing and squalid cities where the altered landscape acted as a visual image of change. In the two decades after 1831 London's population grew from just under two million to over two and a half million. The new industrial cities in the north of England grew at an even faster rate. Factory life, with its regular hours and incessant machines, gave new meaning to 'time'. The pace of life was no longer directed by sunrise and sunset, seasons and harvest, but by the ticking of a clock.

Philosophy, too, felt the changes. The earlier philosophers of the Enlightenment saw the universe as a giant machine: each revolving planet was a cog in the regular movement of the cosmos. In contrast, the new 'Nature Philosophers', prominent in Germany, preferred to see nature as an organism. Not only did this entail a more sympathetic approach to the natural world, it also implied that as an organism it could grow. Accordingly there were grand descriptions of the development of the universe, much as popular books today trace the wonder of the creation of the cosmos from an original 'big bang'.

In the organic sphere the development hypotheses could not be matched by evidence. In the history of kings and queens and the affairs of men and women there was a clear pattern of evolving cultures. But in the animal world no historian had ever recorded a leopard without his spots. The mummified animals of ancient Egypt appeared to be no different to those currently living. So, the evidence clearly pointed to a once-for-all creation. And once created, nothing had changed. From our vantage point 200 years later we may consider the early-nineteenth-century naturalists slow-witted for not reading the message of the rocks and fossils. Surely the layers of rock deposits, each with its characteristic fossils, were clear indicators of the past? But, as has already been pointed out, geology was only an infant science and not yet able to speak.

All too readily we scorn as obscurantist those who opposed new ideas, and heap honour on those who started on the road to our present scientific understanding. Yet the weight of scientific understanding, as then understood, rested with the traditionalists. A 'great lives' approach to the history of science is dangerous. Some important Victorian scientists erred wisely on the side of caution, yet their contribution may be forgotten in the attempt to trace the origin of theories we now believe to be true. In any conflict between speculation and facts, the true scientist must side with the facts. Thomas Huxley in later years described his outlook just prior to the publication of *Origin of Species* as critical of untested ideas:

> 'That which we were looking for, and could not find, was a hypothesis respecting the origin of known organic forms which assumed the operation of known causes such as could be proved to be actually at work. We wanted, not to pin our faith to that or any other speculation, but to

get hold of clear and definite conceptions which could be brought face to face with facts and have their validity tested.'

Chevalier de Lamarck

Jean Baptiste Pierre Antoine de Monet, Chevalier de Lamarck (1744-1829), was, as his long name indicates, a French aristocrat by birth. But he was an impoverished one. As a young man Lamarck served in the army, and when invalided out he turned to studying medicine and biology. At the age of forty-four he was appointed to a position at the Jardin du Roi in Paris, and six years later to a professor's chair at the great Musée d'Histoire Naturelle. This was not a job with much status and he

Much nineteenth-century English architecture harked back to earlier periods of history. St Pancras Station's Gothic arches are seen in this painting. In people's view of history, age followed age in a steady pattern of progress.

The Chevalier de Lamarck was one of the first biologists to come out firmly for the evolution of species. But his explanation of how evolution happened was later to be proved wrong.

had the task of lecturing on Linnaeus' 'insects and worms' — a radical departure from his botanical interests. Although always overshadowed by the more influential Georges Cuvier, then dominating French science, Lamarck was able to publish useful work on taxonomy (biological classification). It is to him we owe the word 'invertebrate', and he coined 'biology' to refer to the study of all living things. When in 1829 Lamarck died, blind and poor, Cuvier denigrated his ideas on evolution claiming they were only fit to amuse the imagination of poets. Not surprisingly his work was therefore neglected, misunderstood, even derided.

Lamarck published his first full statement of evolutionary views in *Philosophie Zoologique* in 1809, the year of Charles Darwin's birth. In an earlier work he had mentioned the idea of evolution, but the 1809 work and those that followed laid down 'laws' of development and gave numerous examples from the animal kingdom. His central thesis was that the structure of animals had changed over time in response to changes in the environment and habitat. A cold autumn day leads us to get out our winter coats, and in so doing we are reacting directly to the environment. So, said Lamarck, the need to pass through confined spaces has elongated the snake's body, and the giraffe has developed a long neck by constantly straining upwards to gather food from trees. Small changes occur during the lifetime of an animal and these changes are passed on to its offspring. The sum of such changes, added over many generations, culminates in a new species.

Research in this century has shown this to be wrong. Changes such as Lamarck describes — changes that have accrued during the lifetime of an individual animal — cannot be passed on from parent to child. By constant exertion a giraffe may lengthen its neck by a fraction, but this extra height will not be passed on to its family. Characteristics acquired during an animal's lifetime are not hereditary, otherwise racehorses would always be sent to stud at the peak of their racing careers instead of on retirement. The aim would be to catch the maximum speed and fitness developed by the animal before it is lost through old age. Yet if Lamarck is wrong, and the acquired muscles and skill are not passed on to the next generation, why do we only breed from thoroughbreds to produce racehorses? Why not any sufficiently fertile nag? Of course, there is such a thing as genetic inheritance — we do inherit something from the parental stock. But, as we now know, it is the genetic make-up fixed at conception, and not the skills and habits acquired during life.

Nevertheless, in the absence of alternative theories, evolution by inheritance of acquired characteristics was a persuasive idea. It persuaded many, Charles Darwin included. Had Lamarck limited himself to this concept, his reputation in the nineteenth century might have been greater. However, in the decades following 1809 and *Philosophie Zoologique* he wrote of an innate power conferred on nature by God. Just as we joke that a ball of string has an uncanny knack of becoming tangled, so Lamarck believed that organisms had a built-in tendency to become increasingly complex until

The iron-smelting works of Coalbrookdale in the English Midlands were archetypes of the Industrial Revolution. These great processes inclined people to see the whole creation as a vast machine.

the pinnacle of creation was reached in the human species. Within each creature there was an inner force which operated continuously for the improvement of the species.

The ancient idea that animals could be naturally but spontaneously generated, that mice were 'created' in dirty shirts and maggots in putrid meat, was seriously questioned some 100 years before Lamarck. Yet at the very lowest end of the scale, the possibility still remained of living organisms arising directly from non-living materials. Lamarck believed in the ancient doctrine of the Great Chain of Being, that spectrum of organisms from the simplest to the most complex. But for Lamarck the spectrum was in motion, with each organism moving upwards to the next level in the chain. C.C. Gillispie has suggested the attractive metaphor of an 'escalator of being'. At the foot of the moving staircase spontaneous generation fed in simple unicellular organisms, and these gradually rose over thousands of years to become

human beings at the top.

Lamarck also spoke of the higher animals having an inner disposition to adapt themselves to environmental changes. This combination of an 'escalator model' and an inner striving did not commend itself to other naturalists. Although Lamarck believed the driving force of the organic escalator to be God, he was denying the instantaneous creation of the higher animals as described in the first chapters of Genesis. He was substituting an almost natural creation among the lower organisms plus a long time period for the higher animals to evolve.

But while there was no proof, Lamarck was dubbed by the English 'the French atheist', and his eager support of the French revolution only confirmed this English distaste. They preferred miraculous creation by the Almighty to spontaneous generation among the protozoa. Rather than invoke a 'universal creative principle' always and everywhere urging creatures to greater complexity, they put their faith in the intervening hand of God. God had populated each level of the staircase of being. The highest animals were created as directly as the lowest; there was no changing of one into the other.

Religion, politics and science here mix — as they always do. The Lamarckian philosophy was rejected not only because the facts weighed against it but because such an outlook was believed to be at variance with cherished political and religious beliefs. The French revolution had brought political instability and antagonism towards the established church; the English would not countenance a son of that revolution telling them that change was the very essence of life.

There is a simple view of scientific endeavour which believes facts and truth to be evident and paramount. The scientist's rule,

as Thomas Huxley expressed it, is 'sit down before fact as a little child, be prepared to give up every preconceived notion, follow humbly wherever and to whatever abyss nature leads'. Yet our perception of truth, our weighing of the evidence, can be dependent on our social and philosophical predilections. The new biological theories held that the lowly could become great. This was seen as a threat by a class-ridden society.

Mrs Alexander's children's hymn of 1848 betrays her Victorian natural theology and social attitudes:

'All things bright and beautiful,
All creatures great and small,
All things wise and wonderful,
The Lord God made them all.

Just as popular astronomy books today rhapsodize over the first minutes of the universe, so the nineteenth-century imagination was captured by accounts of how the cosmos had developed.

The rich man in his castle,
The poor man at his gate,
God made them, high or lowly,
And ordered their estate.'

A hypothesis such as Lamarck's that minimized miracles seemed to attack popular evidence for God. Later in the nineteenth century, after the concept of evolution was accepted and the Revolution but a dim memory, it was realized that Lamarck's theories could be 'deified'. His creative principle was acknowledged as the immanent hand of God working quietly within nature to fulfil the plan of organic development. God was cranking the escalator of being. So for his wisdom Lamarck was reinstated as a scientist. But at the start of the century his opponent Cuvier preached a

How does a giraffe get its long neck? Lamarck believed that the imperceptible stretching caused by reaching for high leaves was passed on to successive generations.

commitment to the fixity of species, not their development, and the poor professor did not live to see his fame.

Mr Vestiges

In 1844 a Scottish publisher, Robert Chambers (1802-71), tried to advocate evolution, Lamarckian-style, to the British public. Sadly his amateurish knowledge and credulous speculations only damaged the evolutionary cause still further. The book, *Vestiges of the Natural History of Creation*, was published anonymously. Chambers wanted to protect himself and his profitable business (to which we owe the fine Chambers Encyclopedias) from those vehemently opposed to any transformist notions. The anonymity also served to add mystery — and hence good publicity — to the book, and in total twelve editions were issued with increasing rumours that the author was none other than Prince Albert. The beans were only spilt after

Chambers' death in a preface to the last edition written by a friend.

Vestiges ascribed the origin of life to a 'chemico-electric process' and suggested we were descended from a type of frog now extinct. The heroine of Disraeli's *Tancred* (1847) exclaims:

'You know, all is development. The principle is perpetually going on. First there was nothing, then there was something; then — I forget the next — I think there were shells, then fishes; then we came — let me see — did we come next? Never mind that; we came at last. And at the next change there will be something very superior to us — something with wings. Ah! that's it: we were fishes, and I believe we shall be crows.'

Vestiges owed its pedigree as much to the German school of Nature Philosophy as to Lamarck. Indeed, *Vestiges* spoke harshly of Lamarck since the French scientist had ignored certain factors which *Vestiges* deemed important. Scotland had close links, at that time, with the Continent and Chambers' writings pulsed with the grand cosmic evolutionism then so popular in Germany and propagated in France by Geoffroy Saint Hilaire. But there were Lamarckian passages on inner forces leading

The social upheaval of the French Revolution was repugnant to the English, who preferred to think that God has set all beings in their immutable stations.

to the modification of body structures and organic complexity. Dominant was a belief that external conditions might directly alter foetal development. A small change at such a crucial moment of development might lead to changes which could accumulate into a new species. No doubt his interest in such foetal changes was occasioned by his own 'alteration': both he and his brother were perfect hexadactyls, each having six fingers on each hand.

Central to the book was Chambers' own commitment to scientific law. The book opened by discussing the nebular hypothesis for the formation of the solar system, an idea which had been put forward earlier in the century by Pierre Laplace. If, Chambers argued, the solar system evolved by means of natural law then it was reasonable to expect evolution by law among organic creations. This did not remove God, rather it credited him with greater glory. For Chambers it was a more noble idea to believe in a God who created laws which in turn acted as his agents of creation, than it was to imagine a meddling creator who stoops 'at one time to produce Zoophytes, another time to add a few marine mollusks!'. The historian Michael Ruse sees behind this enthusiasm for natural law the newly installed printing presses at Chambers' publishing house. Captivated by the new steam technology, Chambers thought of God as the Supreme Engineer.

The scientists and critics lambasted the book and showed up its flimsy foundations. Again, too much speculation compared to fact. Huxley later admitted he had reviewed it with 'needless savagery', yet the attacks only increased the book's popularity and the glamour of the then unknown 'Mr Vestiges'. Charles Darwin was grateful for its publication. By this time he had formulated his own private views on evolution. *Vestiges* might have lacked scientific worth, but it did bring the subject before the British public. Darwin also hoped it would take the main fire of opposition, clearing the ground for his own work.

But Britain was not yet ready for Darwin. In the first half of the nineteenth century the

The typical Victorian family had a strong sense of where it belonged in the social hierarchy. Even their family prayers and their church structures reinforced their sense of an ordered pattern in which all should keep to their allotted places.

dominant belief was in the fixity of species. In his autobiography Darwin comments:

> 'It has sometimes been said that the success of the *Origin* proved "that the subject was in the air" or "that men's minds were prepared for it". I do not think this is strictly true, for I occasionally sounded not a few naturalists, and never happened to come across a single one who seemed to doubt the permanence of species ... What I believe was strictly true is that innumerable well-observed facts were stored in the minds of naturalists ready to take their proper places, as soon as any theory which would receive them was sufficiently explained.'

Three embryos, each of a different species, look quite similar. The writer of 'Vestiges' believed that a series of quite small changes in factors affecting embryos could lead to the development of a new species.

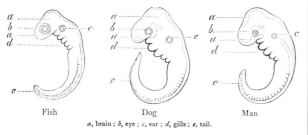

Fish   Dog   Man

a, brain ; *b*, eye ; *c*, ear ; *d*, gills ; *e*, tail.

Even his own correspondence shows that Darwin knew of those who doubted the fixity of species. But we will let that pass. It is indicative that following the furore over *Vestiges* the next decade produced nothing more than idle gossip on the species question.

In London the leading scientist was Richard Owen, and in Paris Georges Cuvier, the first being an ardent disciple of the second. In opposition to Lamarck, Cuvier held that there had been one original, and perhaps divine, creation of animal forms. Variations down the ages were possible, but complete changes from one species to another were not. The vast changes detailed in the fossil record were not reflections of evolutionary development but the history of a series of catastrophes. These catastrophes — floods, earthquakes, climatic changes and the like — had overtaken the earth, destroying all the animals in various locations at various times. Following each catastrophe different creatures had repopulated the stricken zone leaving in the fossil

Richard Owen, the leading British biologist, followed Parisian Georges Cuvier in rejecting the possibility that new species arise from existing ones.

record a discontinuity in animal forms. In the hands of others these 'catastrophes' were soon given a religious veneer by claiming that the repopulation was not the result of natural causes, such as animal migration from neighbouring areas, but divine creation. To English clerics such as William Buckland, Cuvier's earthquakes and floods became divine interventions of which the last was the Deluge of Noah.

Progression, but without evolution

An orthodox believer of those times was likely to hold that the pattern of fossil remains was the result of an alternating sequence of planned destruction and recreation. Following each catastrophe there had been a new creation, but at a more advanced stage *en route* to mankind. This was not wilful misreading of the geological record to support religion. To many the rocks spoke eloquently of Moses' accuracy as a historian. The Flood was described in the Bible and engraved in the rocks. And this was fitting, because their view of biblical revelation allowed them to place the testimony of Genesis on a par with the latest geological research.

In 1823 Buckland discovered in a Yorkshire cave the deposited remains of hyenas mingled with those whose flesh they had eaten, namely lions, tigers and elephants. The news caused a public sensation. As obvious as it was to a Yorkshireman that elephants no longer roamed the Dales, so it was clear that these were the remains of animals caught in the Deluge.

Here, then, was a belief in progression, but without any hint that one animal had been 'transmuted' into another by gradual evolution. As the fossils clearly showed, the demarcation lines were abrupt: one world with its flora and fauna ended, another (more advanced) began. All previous worlds were but preparations for the summit of creation in mankind. Moses was right.

The opposition to this mixture of progression-yet-fixity of species was not evolution

(for that option was still to come) but Lyell's doctrine of uniformitarianism. We have encountered Lyell's geological views in the previous chapter. In biology his outlook was the same: past conditions must be explainable in terms of laws and effects visible in the present. There may be localized change, but the overall effect is constant, much as a kaleidoscope is a constant rearrangement of the same colours only to repeat itself in time.

Lyell argued that the fossil picture of progressive complexity was only apparent. He rejected the idea of an intervening, miracle-working God, pointing out that since the fossil record was hopelessly incomplete a better interpretation of the 'sudden' appearance of particular fossils might be natural animal migration. This was not simply following Cuvier. The French scientist believed in a series of catastrophes beyond human imagination, whether divine in origin or not. Lyell stressed regular, natural processes; there were no forces at work which were not observable in the present.

In the end a transmutational or evolutionary hypothesis had to borrow from both: the gradual naturalism of Lyell and the evidence of progression from Cuvier. Up until Darwin's synthesis of the two (progression through a natural process of evolution) the majority of scientists in both Europe and America felt they had to choose: they could accept either the uniformity of natural agencies or the idea of organic progression. But not both.

Yet both were required for a theory such as Darwin's. He needed to be able to stress the role of ordinary natural agencies (and here he agreed with Lyell), yet see these agencies as producing direction or progression (and here he disagreed with him). For this reason, contrary to some popular understandings of this period, there is no clear-cut history of the new theories being welcomed by the scientists but hindered by the church. A religious conservative such as Sedgwick was only too happy to see progression in the fossil record, while Lyell disagreed. (Lyell was in error: 1-0 to Sedgwick.) But it was Lyell who held the key idea of large-scale effects being accomplished by the accumulation of much smaller effects rather than the impact of a single miracle. (Match to Lyell). Darwin always said that his work 'half came out of Lyell's brain', but he also had a debt to those who stressed the progression seen within the fossil record.

In part it was a religious quarrel: the advocates of a miracle-working God versus those who saw God working through 'natural' causes. It was also an argument over scientific methodology: do you analyze natural history according to present-day phenomena or can you invoke past creative forces, totally beyond any known powers, as part of the explanation?

Behind both progressionism and uniformitarianism lay the firm notion of fixity of species. If the fossils showed increasing complexity of animal forms this was not a record of one form evolving into another, but either a record of successive divine creations or incomplete snapshots through time of animal migrations. Either way, species did not change beyond small local variations. For an explanation of why this belief was so central we must again turn to Georges Cuvier.

Georges Cuvier

Lamarck started with the advantage of both birth and title; Cuvier was an 'outsider' from Jura educated at Stuttgart. Yet where Lamarck seemed destined to fail, Cuvier entered on a career of unchecked success rising to become an official parliamentary spokesman in post-revolution France. He dominated French intellectual life and his ideas, notably his rejection of evolution, enveloped French science for many years. It would be easy to see his work as a hindrance to the emerging evolutionary thinking, especially since his influence extended beyond France to mould scientists such as Owen, the leading British biologist in the 1840s and 50s. But just as Linnaeus' great systematizing work was essential in order to understand the breadth of creation, so Cuvier's

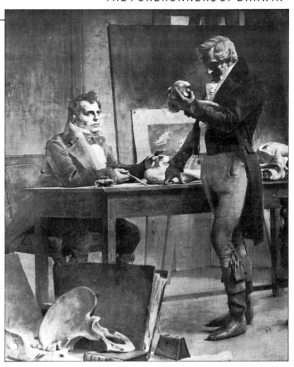

Georges Cuvier developed a system of classification which took account of how species were fitted to their different habitats.

breakdown of the Swede's rigid approach was vital for biologists to appreciate the natural branches of that creation. Cuvier passionately believed in the fixity of species, yet his work produced a vital tool that in the hands of others led to the destruction of this static picture.

Linnaeus, you will remember, classified living creatures into groups dependent on their observed characteristics. Quadrupeds, birds, amphibia, fish, insects and worms were his main groupings. The catalogue was static, much as a bird-watcher's book simply catalogues existing species. Each organism was 'pigeon-holed' in its place in the divine plan of creation. Cuvier, too, believed in an overall plan, but he described animals according to how they functioned in their environment, not on the existence or absence of some arbitrary characteristic. The details of his four main groupings or *embranchements* need not detain us. They were the types of nervous system observable in nature. And having divided living creatures into Vertebrata, Mollusca, Articulata and Radiata, he insisted these were absolute definitions. Nothing that belonged to the animal kingdom could fail to find its place in one of these great divisions.

The significant point is that just as the design of a car differs fundamentally from that of an aeroplane or ship (for they are designed for different media), so Cuvier's *embranchements* represented the different ways organisms might meet the conditions of existence. This approach underlined the stability or fixity of species, for a departure from the basic designs, whether in a spider or a mammal, would render the organism unviable — it would die. In our example the removal of an aeroplane's wings in a misguided attempt to make it look like a ship would simply cause it to crash! The intermediate forms necessary on any evolutionary hypothesis are deemed impossible, just as taking two wheels off a car creates, not a new design of motorbike, but an unstable car!

Cuvier's classification of animals in relation to their viability in the environment thus stressed the dependence of the organisms on their surroundings. His was not a classification based on some arbitrary characteristic: it took into account how the organism functioned. Therefore, so later evolutionists were to claim, if the habitats changed, so must the design of the organism. As a climate gets colder then those animals with fur will survive, while others may not. It was Darwin's belief that altered habitats weeded out unfit specimens and encouraged new strains that had adapted to the new conditions. By itself this belief was nothing new. Everyone recognized the existence of variations, and that offspring always varied slightly from their parents. Lyell was clear that environmental changes favoured some variations of organic structure and destroyed others. But for Cuvier, Owen and Lyell this 'filtering' by the environment was conservative. Small variations were permissible, but an animal born with a structure departing significantly from

its parents would not be able to survive. It simply would not function. Darwin dared to see this adaptation to the environment as a creative force: given a succession of small variations there was no limit to the final change that could occur. One species might evolve into another.

The Linnaean approach was also static in that it attempted to map out the whole of creation. In the bird-watcher's book an attempt is made to index every specimen. Once everything is labelled and pigeon-holed the whole complete interlocking system is gloriously revealed. Cuvier's approach, however, stressed only design. Further designs (or minor modifications along the lines of a stable tricycle derived from a bicycle) always remained possibilities, even though unrepresented in the present animal kingdom. Though denied by Cuvier, the design approach could be used to show how one design could evolve from another.

When Darwin came to set sail on HMS *Beagle* in December 1831, the doctrine of the fixity of species was as certain in his mind as it was in any other naturalist's of his day. There had been speculation on evolution, but there was no proof and no cogent mechanism to show how such changes might occur. The new palaeontologists had unearthed fossils of organism now believed to be extinct, and marvelled at the progressive plan leading up to mankind. The systematists described an increasingly large and complex biological world. Worryingly they had noted hybrids that appeared to be new species, but although variations were acknowledged the fundamental axiom was that organic types were distinct and eternal. The pressure from the environment ensured that organisms significantly departing from the master plans would not survive. Such a belief was bolstered by politics and religion; it was convenient to both. And the young Darwin, intent on returning home from his voyage to take up a vocation as a Church of England clergyman, scarcely believed otherwise.

4
ORIGIN
OF SPECIES

Charles Darwin was born on 12 February 1809. His father, Robert, was a Shrewsbury doctor, and Charles was the youngest but one in a family of two boys and four girls.

Sent to the local Grammar School, Charles summed up the teaching he received as strictly classical and 'as a means of education . . . simply a blank'. He enjoyed reading poetry, and towards the end of his school days earned

Charles Darwin's interests as an undergraduate were not primarily academic. He was devoted to hunting and shooting.

himself the nickname 'Gas' for spending many hours conducting chemical experiments in the tool-house under the direction of his elder brother. He also found pleasure in watching the habits of birds and wondered 'why every gentleman did not become an ornithologist'.

Caring for nothing but shooting

'Early in my school days a boy had a copy of the *Wonders of the World*, which I often read and disputed with other boys about

the veracity of some of the statements; and I believe this book first gave me a wish to travel in remote countries which was ultimately fulfilled by the voyage of the *Beagle*. In the latter part of my school life I became passionately fond of shooting . . .'
Autobiography, Charles Darwin

Charles may have been passionate about shooting, but it was not an enthusiasm that touched his more routine studies. His father shared with the school headmaster a low view of Charles' attainments and removed him at sixteen years of age, telling him:

'You care for nothing but shooting, dogs and rat-catching, and you will be a disgrace to yourself and all your family.'

Hoping for better things, Dr Darwin packed his son off to Edinburgh University to train for a career in medicine. His grandfather Erasmus had been a doctor so the profession ran in the family. His father hoped (and to a degree believed) Charles would make a successful physician. But Charles found the lectures dull, and passed his time in the company of other young men interested in natural history. At times, too, his medical studies were distasteful. The discovery of chloroform (by the German chemist Justus von Liebig) was still six years away and the experience of witnessing operations without anaesthetic haunted Darwin's sensitive mind. After only two university sessions his father perceived that medicine was not, after all, his son's vocation and proposed that Charles should become a clergyman.

'He was very properly vehement against my turning an idle sporting man, which then seemed my probable destination. I asked for some time to consider, as from what little I had heard and thought on the subject I had scruples about declaring my belief in all the dogmas of the Church of England; though otherwise I liked the thought of being a country clergyman. Accordingly I

Darwin's medical studies at Edinburgh University were another blind alley. His eventual career had its origins in his spare-time hobby of natural studies.

read with care Pearson on the Creeds and a few other books on divinity; as I did not then in the least doubt the strict and literal truth of every word in the Bible, I soon persuaded myself that our Creed must be fully accepted.'
Autobiography

Looking back Darwin himself records how ironic it was that he once intended to be a clergyman in view of the questions he raised for the Victorian church. And indeed he never completed his selected path; his father was again to be disappointed. But in Charles' own words, 'during the three years which I spent at Cambridge my time was wasted'. From the perspective of a clerical education Darwin was correct. Though he gained a passable degree his attendance at many lectures was

'When at Cambridge I used to practise throwing up my gun to my shoulder before a looking glass to see that I threw it up straight. Another and better plan was to get a friend to wave about a lighted candle and then to fire at it with a cap on the nipple, and if the aim was accurate the little puff of air would blow out the candle. The explosion of the cap caused a sharp crack, and I was told that the Tutor of the College remarked, "What an extraordinary thing it is, Mr. Darwin seems to spend hours in cracking a horsewhip in his room . . . "

'From my passion for shooting and for hunting and when this failed for riding across country I got into a sporting set, including some dissipated, low-minded young men. We used often to dine together in the evening, though these diners often included men of a higher stamp, and we sometimes drank too much, with jolly singing and playing at cards afterwards. I know that I ought to feel ashamed of days and evenings thus spent, but as some of my friends were very pleasant, and we were all in the highest spirits, I cannot help looking back to these times with much pleasure.'

only nominal and his interest even less. Instead Charles collected beetles, and even captured a rare specimen which was later illustrated in Stephen's *Illustrations of British Insects*.

'No pursuit at Cambridge was followed with nearly so much eagerness or gave me so much pleasure as collecting beetles. I will give proof of my zeal: one day on tearing off some old bark, I saw two rare beetles and seized one in each hand; then I saw a third and new kind, which I could not bear to lose, so that I popped the one which I held in my right hand into my mouth.'
Autobiography

Reaction was prompt. The outraged beetle immediately squirted a burning and unpleasant

At Cambridge Darwin made a detailed study of beetles, provoking a friend to caricature him so engaged.

As a young man Charles Darwin developed the techniques and interests of a naturalist, even though this was not the subject of his academic studies.

liquid into Darwin's mouth and the eager collector was forced to spit it out.

In retrospect this beetle mania was no mere hobby. He used his years at Cambridge to develop skills in shooting and stuffing birds, collecting insects, and surveying rocks. He also read the lore of worldwide travellers such as his hero Alexander van Humboldt. In much of this he was encouraged and assisted by his professors, and the outcome was a repertoire of techniques essential to a working scientist — techniques which would stand him in good stead

John Henslow was the guide of Darwin's developing biological career. He was to be deeply disappointed when his friend's evolutionary theories did not find an explicit place for God's activity.

during his own explorations around the world.

Through his interest in natural history he struck up a friendship with John Henslow, the professor of botany. Darwin was known as 'the man who walks with Henslow', and when Darwin graduated in the summer of 1831 and returned to Shrewsbury it was Henslow who provided the next step. The professor wrote to his protegé of an intended voyage to Tierra del Fuego and then home via the East Indies. The master of the vessel, Captain Robert FitzRoy RN, had asked for a naturalist and a companion on his hydrographic survey of South America. The request, after others had refused, had been passed to Henslow, and thence to Charles. Would Charles like to take it up? Darwin jumped at the offer. The Humboldt narratives had given him a foretaste of foreign travel, and his studies in natural philosophy, albeit limited, had left him with 'a burning zeal to add even the most humble contribution to the noble structure of Natural Science'.

Hating every wave

HMS *Beagle* sailed out of Plymouth two days after Christmas in 1831. In the five years of the voyage Darwin never mastered his sea-sickness and came 'to hate every wave of the ocean'. At

times too he was homesick. Yet the voyage made his name.

At ports he investigated the terrain and captured the animals. He studied the rocks and made geological reports. The flora and fauna he collected, classified and, if possible, shipped back to a waiting Henslow. He worked almost without ceasing, filling over 1,300 pages with geological notes, 24 notebooks with impressions, 368 pages with zoology, and a diary of nearly 800 pages. There were islands on which (it seemed) he collected every creature that flew, ran or crawled. But, contrary to popular opinion, he did not stumble across a new theory of evolution. All his industry provided the fuel for the reflection and theorizing for which he was to have opportunity later on. The reputation that awaited him when he returned to London in October 1836 was built on the quality of his descriptive work and his contribution to geology.

On leaving England Darwin entertained no notions of evolution. He believed that species were fixed, as ordained by their creator. But he did not believe this with the vehemence that his cabin companion FitzRoy was later to maintain. At this time Darwin held his scientific views much as he did his religion — simply as part of his intellectual and social heritage. He was an English gentleman. The Captain was also a man interested in the sciences, but whereas over the five wave-tossed years Darwin began

▶ p66

The model for this drawing by Charles Darwin was a vampire bat caught on the back of his horse in South America.

DARWIN'S VOYAGE ON THE BEAGLE

'The voyage of the *Beagle* has been by far the most important event in my life, and has determined my whole career . . . Everything about which I thought or read was made to bear directly on what I had seen or was likely to see; and this habit of mind was continued during the five years of the voyage. I feel sure that this was a training which has enabled me to do whatever I have done in science.'
Charles Darwin

The sloths and armadillos Darwin found in South America were similar to extinct species whose fossils had been unearthed.

HMS Beagle's five-year voyage took her right round the world. The opportunity this gave the young Darwin to develop his natural studies was seized with eager hands.

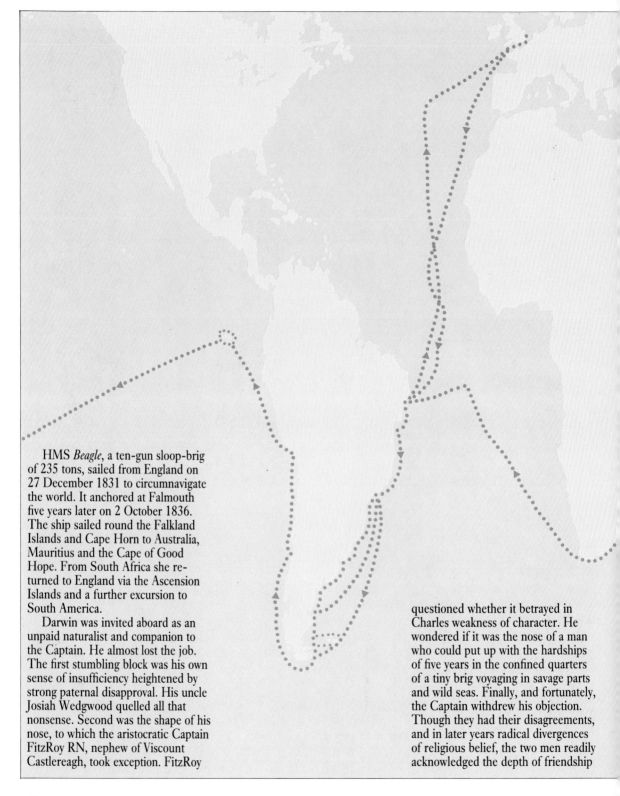

HMS *Beagle*, a ten-gun sloop-brig of 235 tons, sailed from England on 27 December 1831 to circumnavigate the world. It anchored at Falmouth five years later on 2 October 1836. The ship sailed round the Falkland Islands and Cape Horn to Australia, Mauritius and the Cape of Good Hope. From South Africa she returned to England via the Ascension Islands and a further excursion to South America.

Darwin was invited aboard as an unpaid naturalist and companion to the Captain. He almost lost the job. The first stumbling block was his own sense of insufficiency heightened by strong paternal disapproval. His uncle Josiah Wedgwood quelled all that nonsense. Second was the shape of his nose, to which the aristocratic Captain FitzRoy RN, nephew of Viscount Castlereagh, took exception. FitzRoy questioned whether it betrayed in Charles weakness of character. He wondered if it was the nose of a man who could put up with the hardships of five years in the confined quarters of a tiny brig voyaging in savage parts and wild seas. Finally, and fortunately, the Captain withdrew his objection. Though they had their disagreements, and in later years radical divergences of religious belief, the two men readily acknowledged the depth of friendship

Captain FitzRoy sailed the Beagle to the tip of South America to restore to their homeland three Fuegians who had been taken hostage on an earlier voyage.

forged on the voyage. 'Darwin,' FitzRoy recorded, 'is a regular trump!'

The object of the *Beagle* voyage was briefly stated by Charles:

'To complete the survey of Patagonia and Tierra del Fuego, commenced under Captain King in 1826 to 1830; to survey the shores of Chile, Peru, and some islands in the Pacific; and to carry a chain of chronometrical measurements round the world.'

Darwin's travel journal, originally published as the third volume of the official *Narrative of the Surveying Voyages of HMS Adventure and Beagle* in 1839, is one of the best travel books ever written. His publisher soon realized this and separate editions followed in 1845 and 1860. In Victorian England it fed a constant clamour for travelogues and natural history. Even today it is very readable and modern editions are still published.

Darwin entertained his readers to his observations on the geology and animal inhabitants of lands stretching from America to Australasia. He described active volcanoes and long-petrified forests, animals both living and extinct. On the coast south of Buenos Aires he came across a tomb of ancient giants. From within an area of 200 metres he dug out the remains of nine large quadrupeds including a *Megatherium* or giant sloth and an extinct elephant later named *Mylodon darwinii*. Here Darwin was confronted with voices from the past, voices which were echoed in the still-existing sloths and armadillos seen on his inland expeditions. If Cuvier was right that such ancient worlds were destroyed only to be replaced by new creations, it seemed strange that the same design (albeit made to smaller dimensions) should be repeated from one creation to the next.

But it was not all natural history. Geology was for long Darwin's prime interest, and on reaching the Keeling

The Galapagos Islands, with their profusion of distinct species, were a natural place for questions to arise about the reasons for variations between species.

Islands he was fascinated with the vast coral reefs. He was to describe his theories as to their formation in his book *The Structure and Distribution of Coral Reefs* (1842). In Chile he witnessed what was considered the worst earthquake ever to have occurred in that land. He was horrified by the damage caused, yet captivated by the power unleashed. He realized how such earthquakes could transform the terrain of any country. He became a convert to Lyell's teaching on the sufficiency of gradual effects rather than to sudden divine interventions.

Captain FitzRoy went to Tierra del Fuego at the tip of South America, with a purpose other than surveying. On the previous voyage he had taken three native Fuegans hostage in reprisal for the theft of the ship's

Not all Darwin's studies dealt with land species. He became absorbed for a while in the study of how coral formed.

whale-boat and a fourth had been bought from his parents for a pearl-button. On arriving in England one died of smallpox, but the others were given (at the Captain's own expense) a crash course in British civilization and Christianity, and paraded in front of the Queen. FitzRoy's hope was to reintroduce Fuegia Basket, York Minster and Jemmy Button (as they were named) to their own people as missionaries. To that end a clergyman, the Reverend Richard Matthews, was included in the ship's company, as he intended to live amongst the Fuegan tribes. The attempt was a failure. Matthews soon lost all his equipment and had to be taken back on board. The three Fuegans remained but imparted only a little of the English language to their fellow tribesmen. The hoped-for propagation of British manners and customs was doomed from the start! Darwin concludes:

'Everyone must sincerely hope that Captain FitzRoy's noble hope may be fulfilled, of being rewarded for the many generous sacrifices which he made for these Fuegans, by some shipwrecked sailor being protected by the descendants of Jemmy Button and his tribe!'

Fortunately for HMS *Beagle* no such shipwrecking ruined the returning voyage. Other naturalists like Edward Blyth lost all their valuable specimens through shipwreck, but Darwin was lucky. He returned with his precious cargo and notes intact. As a scientist his reputation was assured.

On his arrival at home in Shrewsbury, his father turned to Charles' sisters and exclaimed, 'Why, the shape of his head is quite altered!' The school-boy failure had returned, and as a man he had matured. He had found a vocation.

to question the immutability of species, FitzRoy became increasingly certain of the truth of Noah's Flood. When the story of the *Beagle* was published FitzRoy added a chapter on how the voyage had confirmed his literal understanding of the Bible.

For study on the voyage, Darwin had obtained the first volume of Lyell's *Principles of Geology*. Lyell, as we have seen, undermined the views of the orthodox by invoking purely natural causes and endless time. Henslow had advised Darwin to read the book, yet not to believe it. But Darwin was enthralled. It opened up a new way of understanding the terrain of the countries he visited. In matters scientific he found his cabin-mate in the printed wisdom of Lyell, not the fervent religion of FitzRoy. The second volume of *Principles* reached Charles in Montevideo the following year. In the opening chapters Lyell outlined the evolutionary views of Lamarck and, although he dismissed these new ideas in favour of a constant ebb and flow of animal migrations, Lyell no doubt sowed questions on the fixity of species in the young Darwin's mind.

Other naturalists had reflected long on the breadth of God's creation. Yet, in his voyage around the world, Darwin was in a unique position to witness at first hand the vast numbers of species. The sheer magnitude of the variety made him wonder whether differences between one animal and the next should be ascribed to an over-generous God or to some more natural cause. The animals of Australia — the kangaroo, the emu, the duck-billed platypus — were so different to animals in the rest of the world that he pondered whether two distinct creators had been at work. In the Galapagos Islands off South America he was told that the finches and tortoises differed marginally from one island to another. But why should God create different finches to populate each island in such a small archipelago? Why should the biology of the Galapagos resemble that of mainland South America but not that of the distant Cape Verde Islands, even though the climate and environment were similar? He reflected:

'Seeing this gradation and diversity of structure in one small, intimately related group of birds, one might really fancy that

Following his visit to the Galapagos Islands, Darwin realized that the finches from different islands were not the same. The finches in this photograph come from different islands.

from an original paucity of birds in this archipelago, one species had been taken and modified for different ends.'

The approach to Fernandina, one of the Galapagos Islands in the Pacific Ocean. The islands are now a national park and wildlife sanctuary.

But we must sound a note of caution. The above sentence, now famous in the history of science, dates not from his on-board notebooks but from the second edition of the journal of the *Beagle* voyage published nearly a decade after he had left the Galapagos. By then (1845) his evolutionary theory was well advanced, and the maturer Darwin was slipping in hints of his labours subsequent to his return to England. His notebooks of the time contain no clear questioning of the stability of species, only amazement at its fecundity.

In a scientific theory as bold and comprehensive as the completed *Origin of Species* (1859) there is seldom one single moment of discovery. Stories of the apple falling on Newton's head, or of Archimedes leaping from his bath crying 'Eureka', are just that: good stories. But they are hardly complete history. And the path

to discovering the mechanism of evolutionary change was also complex and long.

As we shall see, in his autobiography Darwin gives October 1838 as the moment of revelation. But before then his mind needed to be prepared to perceive the same data, such as his specimens from the Galapagos, in a new and revolutionary way. If there was an instant of revelation, then there were also long months of speculation and amassing of facts. He needed to go over his notebooks and think afresh on the sights that had puzzled him in the five years of the voyage. Only then could the jigsaw suddenly lock into place. Vital to this new perception were three elements, all garnered (but not assembled) on his five-year voyage around the world.

□ *The superabundance of animal types and the fine gradations between them.* Darwin was saved from

Linnaeus' approach of rigid organic classifications since it was hard for him to decide where one category stopped and the next began. Were the Galapagos finches different God-created species, or were they varieties of one parental stock?

□ *The unique adjustment or adaptation of each organism to its environment.* Darwin visited both mountain and plain, rain forest and barren rock. His university studies in the natural theology of Paley had taught him to note how God had designed each animal for its own peculiar habitat. Darwin indeed noted this, but it seemed the argument could be turned on its head: what if animals naturally developed *in response to* their habitats, instead of being supernaturally formed *for* their habitats? Take, for example, Ray's wonder at the way bees do not swarm until the plums are ripe. Is this divine timing? Or is it because before this moment in summer bees simply cannot survive?

Darwin followed others in studying the relationship of organisms to their environments, bees to flowers, for example. But he came to understand this relationship in a totally new way.

□ *The ever-changing nature of the world in which animals lived.* This was Lyell's influence. The *Principles of Geology* convinced Darwin that the earth was older than 20,000 years, and his own eyes witnessed the power of volcanoes and earthquakes to shape the environment. In Chile he saw how a single earthquake could raise the ground level by almost a metre.

All Darwin needed to do was to assemble these three pieces. Once he abandoned belief in the fixity of species the way was open to see that if the environment gradually changed then those animals that adapted to it would survive. His Galapagos finches were a case in point: some fed on insects, others on seeds. Their different beak shapes allowed them to occupy different ecological niches. But they may have started out

AN UNSOLVED MYSTERY

According to his autobiography Darwin first grasped the importance of natural selection in 1838. He worked on the idea over the following years and wrote down a summary of his arguments in 1844. This essay resembles the final *Origin* of 1859, but it lacks the detailed argument and examples necessary to gain credence in a sceptical Europe. But why did Darwin delay a further fifteen years before publishing his views?

1844 was the year of the anonymous *Vestiges*. Was it fear of the scorn and opprobrium that attached itself to that book which made Darwin delay? Was it sensitivity to the religious feelings of his own dear wife? Was it

simply his desire for perfection, and his intention to write a fully comprehensive account of the new theory?

Exploring the psyche of a man of Darwin's complexity, reading between the lines of his correspondence and notes, has created a Darwiniana industry spawning many books and learned dissertations. But there is still little certainty.

It is with fascination and frustration that we see him turning away from his species work to study barnacles. During his voyage he had collected

Cirrepedia, or barnacles, including one specimen which was so different from the others that it demanded a new sub-order for its classification. Darwin intended to revamp the muddled classifications of former naturalists and anticipated this would take him a year or two. In fact it absorbed eight long years, and although he at first spoke of his 'beloved' barnacles he slowly came to hate them and yearned to be finished with the demanding task in which he had permitted himself to become engrossed — or entrapped.

Just as his 'studies' at Cambridge equipped him for his work aboard the *Beagle*, so his detailed study of barnacles provided a solid background in systematic biology. By the time he had seen his four volumes on *Cirrepedia* through the press, Darwin

Darwin, seen here in a greenhouse at Down House, was no mere theorist. His detailed studies had made him a highly accomplished naturalist.

was a disciplined and trained naturalist. He was also not well. His troubles included headaches, stomach pains, severe fatigue and general indisposition. Some have retrospectively diagnosed brucellosis, others Chagas disease — an infection carried by an Argentinian insect, and Darwin records in his journal how he was once bitten by such a creature. Others have attributed the illness to the forcefulness of his father and the inadequate love received while Charles was still young. His troubles, they claim, sprang from psychological rather than physical sources. From a religious perspective, some modern-day scoffers of evolution have sought to discredit Darwin by seeing in his illness and reluctance to publish the effect of guilt incurred through harbouring a godless doctrine. A doctrine, they claim, Darwin knew to be false.

The solutions proffered may tell us more about the intention of the would-be diagnosticians than the origin of the illness itself. Avoiding the extremes of psychoanalysis, many firmly believe that the illness was mental, although the symptoms were real enough. Darwin had slowly given up his faith; Emma, his dear wife, was never to do so. Apart from causing conflict within his own family, Darwin knew he would not be spared scientific and clerical attack when his views were eventually made known. He needed time to amass overwhelming, and hence convincing, evidence. It was also 'like confessing a murder', and he jotted in an early notebook, 'Mention persecution of Astronomers'.

on the islands from one species which made the journey across from the mainland. As successive generations colonized the islands in the archipelago they adopted different lifestyles and menus according to how their beaks developed. A finch lucky enough to manage a diet not followed by his fellows was in for a feast, and was more likely to survive than if he had to take his share of the limited amounts of regular food. Evolution was nothing more than the accumula-

Thomas Malthus' studies of increasing growth in populations are highly relevant to human welfare today. But for Darwin they offered a clue as to how populations of different species were controlled by processes of selection.

WAS DARWIN THE FIRST?

Was Charles Darwin the first to suggest that evolution works by means of natural selection? As we have already seen it was by then quite common to speculate about evolution, however such speculation might be scorned by English scientists suspicious of French heresies. Neither was the idea of the effect of competition in the wild altogether new. It was clearly outlined in Lyell's books. Yet there it was a conservative force, limiting the amount of change possible within any environment. Darwin achieved the magisterial synthesis in which Lyell's boundaries became paths to species change, and foreign speculation was replaced by amassed empirical facts. But was he the first to accomplish this?

In later editions of *Origin*, Darwin traced some of the historical antecedents to his own theory. Here we find reference to Buffon, Lamarck, Owen and other naturalists too numerous to mention. He also referred to an 1831 book by Patrick Matthew, *Naval Timber and Arboriculture*, in which the Appendix has an outline of evolution based on principles of natural selection. Darwin acknowledged the theory was the same as his own, yet claimed ignorance of its existence until Matthew pointed out the fact some four months after the first edition of *Origin* had been published. Matthew's ideas antedate Darwin by almost thirty years and he insisted on his priority by engraving 'Discoverer of the Principle of Natural Selection' on his visiting cards! Yet he had completely failed to see beyond tree cultivation to the importance of his theory to biology as a whole. He had only written an appendix, not an all-embracing theory of animal origins.

Whether Darwin was in fact ignorant of Matthew's book has been questioned. Loren Eiseley, noted writer on Darwin's era, sees similarity of expression between phrases in Matthew's Appendix and Darwin's trial essay of 1844. More startlingly he highlights earlier published papers by the naturalist Edward Blyth and claims that from these Darwin derived his inspiration, although refusing to acknowledge it.

Blyth, a year younger than Darwin, was born into a poor family. Though naturally a scholar, he only received a trade school education and ran a chemist's shop south of London. Eventually, in 1841, failing health forced him to leave England and he became curator of a museum in Bengal. From there he published numerous contributions to the natural history of South-east Asia. However, it was his papers in *The Magazine of Natural History* of 1835 and 1837 which excited Eiseley's interest. These magazine articles would have reached Darwin during the very years he began first to ponder the species question. Blyth refers to natural selection and although his conclusions match those of Lyell (that is, that selection conserves animal types and discourages evolution) he hints at the possibility that selection might lead to the propagation of new breeds 'very unlike the original type'.

Darwin knew Blyth and in other areas quoted extensively from his work. But there is never any mention of these two key articles. When Eiseley finds Darwin using similar phrases to those used in the articles, his startling conclusion is that Darwin used his colleague's work. And he used it without acknowledgment, as a stepping-stone to his own theory and scientific fame.

tion over many years of countless small changes in beaks, in wings, in feathers. Survival implied change or, put the other way, the mechanism of change or evolution was the struggle to survive.

Confessing a murder

On his return to England in the autumn of 1836 hard work awaited Darwin. Any thoughts of a quiet country vicarage quickly evaporated in the face of boxes and boxes of specimens that needed to be classified. He also had to write up his journal.

Darwin settled in London and married Emma, daughter of Josiah Wedgwood the giant of the porcelain industry. The almost casual letters he had posted back to Henslow from around the world had opened a door into the scientific world in a way that he never anticipated. His university friend had read the letters to the Cambridge Philosophical Society whose printed *Proceedings* had passed to the more prestigious London Geological Society. His mentor Charles Lyell sought him out, and within a few months Darwin was himself a member of that esteemed Society. Two years later he was its secretary. He wrote to a friend that his first visit to London had been passed 'in most exciting dissipation amongst the Dons of science'.

In one way Darwin's natural studies were quite normal. He was one of a great number of enthusiastic students of nature.

Ernst Mayr, in his definitive *Growth of Biological Thought* simply disagrees. Some of Darwin's notebooks, unpublished when Eiseley wrote, prove his originality. In the passages relating to the crucial year of 1838 it is clear that Darwin had not formed his theory of natural selection before reading Malthus. Certainly there are no hints in these private papers that he had derived ideas from Blyth. Darwin had no doubt studied Blyth's papers, but paid no further attention to them as he was by then unenthusiastic about the conservative conclusions they offered. In the main they only repeated Lyell's teaching, and Darwin was moving away from his old mentor's position.

More important was the work of German biologist C.G. Ehrenberg on micro-organisms. One of Ehrenberg's most dramatic findings was the incredible rate of reproduction of such organisms, which seeded Darwin's mind with ideas on the sheer super-fecundity of nature. When, then, he read Malthus on the effects of limited resources among growing human populations, the relation of the human to the general biological world was clear. He saw that a large population of organisms, with each individual differing naturally from the others, would gradually change. Helpful varieties would remain while others, less 'fortunate', would die.

Surprisingly the day of insight, 28 September 1838 did not bring forth a single *Eureka!* or, in the notebooks, even a single exclamation mark! To Darwin it was just one more stage in the theory — a new (and promising) line of approach where others had been blind alleys. It would take weeks to test it as he had other ideas. And there were problems: he knew neither the cause of the variations nor the mechanism of hereditary transmission. A theory of evolution without a base of genetics was a shaky structure.

In the end a scientific theory as wide-sweeping as that of natural selection requires not just one supporting column but many. Not until most of the edifice is in place can the scientist see what he is building. Most of the elements of Darwin's theory (super-fecundity, variations, selective pressure) were widely known before Darwin. It took a man of Darwin's genius to assemble the blocks into a new order and so reveal a structure where others saw only chaos.

In March the following year Darwin met John Gould, the Zoological Society's taxidermist and a leading ornithologist. It was a moment of turning. Gould not only confirmed that the ostrich Charles had found in South America was a new species (and proposed the name *darwinii*), but informed an incredulous Darwin that his Galapagos specimens contained three different species of mocking-bird and thirteen species of finches. Was it then that the old notions of the fixity of species were finally undermined?

Within four months of his talk with Gould, Darwin opened the first of a number of notebooks on 'the species question'. Just over a year later (autumn 1838) the role of natural selection as the key to his theories became clear to him. He had been studying the way pigeon fanciers bred their birds for certain characteristics, carefully selecting which pairs should be used in mating. He recorded in his autobiography:

'I soon perceived that Selection was the keystone of man's success in making useful races of animals and plants. But how selection could be applied to organisms living in a state of nature remained for some time a mystery to me. In October 1838, that is fifteen months after I had begun my systematic enquiry, I happened to read for amusement "Malthus on Population", and being well prepared to appreciate the struggle for existence which everywhere goes on from long-continued observation of the habits of animals and plants, it at once struck me that under these circumstances favourable variations would tend to be preserved and unfavourable ones destroyed. The result of this would be the formation of a new species.'

Nobody today would read Malthus for 'amusement'! The Reverend Thomas Malthus' *Essay on the Principle of Population* was a grim inquiry into the effect of human population growth given limited foodstocks. Populations,

he claimed, grow geometrically, that is, a mother and father may have only two children (and most Victorian parents had far more), but each of these may beget a further two, and each of those a further two, and so on. Food supplies, on the other hand, only increase arithmetically: $1 + 1 + 1 + 1 \ldots$ The result, as population outstrips available resources, is inevitable famine, disease or bitter fighting to obtain the limited food. Consequently, the population will be reduced in line with the available harvests.

The political and moral point of Malthus' writings need not detain us. Darwin realized that if the same theory applied to the animal kingdom then the resulting struggle for meagre resources would ensure only the 'fittest' or best-adapted would survive. The fastest lions would catch the most zebras and their less fleet-of-foot colleagues would die out. Furthermore, since the offspring of the surviving lions would inherit these better genetic characteristics, this increase in speed would be passed on to subsequent generations. The accumulation of a whole series of such improvements might result in a new species.

This filtering effect of a crowded habitat, or natural selection as Darwin termed it, was not a new idea either to biology at large or to Darwin himself. Lamarck, Paley, Humboldt, Lyell, even his grandfather Erasmus, had all mentioned it. But in their hands it was a conservative force, killing off those animals that departed from the God-given norm. Deviations from God's carefully designed forms would spell disadvantage and disaster, not improvement and evolution.

They argued that animals with any significant change from God's optimal design (and you would need large changes if they were to be the start of a new species) would automatically be at a disadvantage. Selective breeding might produce new forms among pigeons or domesticated dogs, but that was because they were protected. They were sheltered from the rigours of competition. And even here there was a limit to the changes which breeders could introduce. In the wild, the jump required to move from

It was to be years after his return from the voyage of HMS Beagle before Darwin's studies were sufficiently advanced for him to put forward his evolutionary theories in a public way.

Darwin's notebooks are full of detailed research which he believed was needed before people would accept his revolutionary ideas.

one viable species to another was simply too great. And, anyway, the whole concept violated the notion that species are eternally distinct.

Darwin observed the same natural world but with a new factor in his mind. He saw the same struggle for existence, but he no longer believed species were distinct entities. And his geology had taught him the power of changes by small degrees. Towering mountain ranges had been gradually built up, section by section, century by century. Provided each variation offered the animal some advantage, however small, over his fellows, then the conservative force became creative as advantageous variations were accrued over countless generations. The origin of species could be laid at the door of variations between parent and child, selection by the environment of those that were advantageous, and endless time for these small changes to add together to make sufficient differences to ensure a new 'species'.

'Here then I had at last got a theory by which to work; but I was so anxious to avoid prejudice, that I determined not for some time to write even the briefest sketch of it. In June 1842 I first allowed myself the satisfaction of writing a very brief abstract of my theory in pencil in 35 pages; and this was enlarged during the summer of 1844 into one of 230 pages, which I had fairly copied out and still possess.'

Here indeed was a theory to work by. But there were still elusive gaps. Why, for instance, did offspring vary from their parents? Using a human analogy, why were not all children like identical twins? And how were such variations, once introduced, transmitted through inheritance? Darwin never solved these mysteries and they remained awkward questions throughout his life. He needed variations to apply to the sieve of natural selection, but while he could highlight their existence he could not account for the source. Nevertheless, just as you can explain to a novice how to drive a car without delving into the chemistry of petrol and air explosions, so Darwin set about tracing how species altered without fully understanding the source of the power.

Ironically an Austrian monk, Gregor Johann Mendel, had the keys to unlock Darwin's puzzle, but though Mendel came to London for the Great Exhibition of 1862, the two men never met. Mendel's scientific papers sank almost without trace until they were rediscovered at the start of the twentieth century. Although Darwin may have put great store by his theory of the survival of the fittest, it was left for later generations to understand the origin of the fittest.

Initially Darwin wrote two sketches of his ideas, one in 1842 and the other in 1844. The earlier one was a mere pencilled summary and was only found among Darwin's papers some years after his death. However, by the New Year of 1844 Charles could write to his friend and confidant Joseph Hooker that at last 'gleams of light have come, and I am almost convinced (quite contrary to the opinion I started with) that species are not (it is like confessing a murder) immutable . . . I think I have found out (here's presumption!) the simple way by which species become exquisitely adapted to various ends.'

An unexpected letter

The 1840s were difficult years in Victorian Britain. Hunger, poverty and rumblings of revolution made the socially disruptive doctrine of evolution unacceptable to the establishment. The anonymous *Vestiges* appeared in 1844. Darwin condemned it along with others:

'The writing and arrangement are certainly admirable, but the geology strikes me as bad and his zoology far worse.'

The time, he reasoned, was not yet right for his own publication, but his views were sufficiently advanced for him to leave instructions to Emma to publish his longer essay in the event of his death.

Why Darwin did not publish immediately remains an enigma. Certainly his arguments needed polishing and supporting with detailed evidence. But Darwin chose to turn his attention away from the species question to the study and classification of barnacles. Illness also sapped his strength and concentration, although it also had an advantage, in that it gave him a reason to withdraw from public and social life and retire to a house at Downe, some sixteen miles south of London. Here he had space to carry out his many botanical experiments and sufficient time to complete his writing, both on barnacles and on the origin of species. Down House, Charles and Emma's home, became so taken over by these fascinating marine creatures that when one of their small sons went to tea with some young friends he solemnly asked when *their* father 'did his barnacles'!

Though illness sometimes made work impossible he eventually wrote to Hooker in 1854:

'I have been frittering away my time . . . sending ten thousand barnacles out of the house all over the world. But I shall now in a day or two begin to look over my old notes on species.'

By the following year he was hard at work, and in April of 1856 revealed to Lyell the stage he had reached. His old friend urged Charles to publish. Darwin admitted he 'certainly should be vexed' if anyone were to publish a comparable theory before him. Yet he dallied for fear his arguments were insufficiently supported by documented evidence. He began writing the following month, but with twenty years of thinking and collecting behind him, he seemed to be looking more for comprehensiveness of evidence than for speed of completion. He planned a massive book, and continued (with Hooker's encouragement) until the summer of 1858.

Darwin sometimes feared the correspondence which awaited him each day. His heart sank if the post was heavy, for his workload and illness made conscientious replies a chore. But the letter Darwin received from the naturalist Alfred Russell Wallace one morning in the summer of 1858 made him sink even lower. It shook him to the core. Another man, with another illness, but on the other side of the world, had beaten him to it. Darwin held in his hands four thousand words entitled *On the Tendency of Varieties to Depart Indefinitely from the Original Type*. It was ready to publish, and Wallace was turning to Darwin in the hope he would steer it towards one of the learned societies or journals.

In dismay Darwin sent the manuscript to Lyell:

'Your words have come true with a vengeance — that I should be forestalled . . . I never saw a more striking coincidence; if Wallace had my MS sketch written out in 1842, he could not have made a better short abstract! Even his terms now stand as heads of my chapters. Please return me

(Wallace's manuscript), which he does not say he wishes me to publish, but I shall, of course, at once write and offer to send it to any journal. So all my originality, whatever it may amount to, will be smashed . . . '

Wallace was fourteen years Darwin's junior. He was an able naturalist yet little known in Europe because he had spent most of his adult life in South America and the Far East. It was while lying sick of malaria at Ternate, an island near New Guinea, that he remembered the ideas of Malthus — ideas that he had read a dozen or so years before and, unknown to him, had also had such an influence on Darwin. Wallace later recalled:

'It occurred to me to ask the question, Why do some die and some live? And the answer

The Irish potato famine of the 1840s was one source of the widespread social ferment which made the middle classes suspicious of any radical theory.

At Down House in Kent Charles Darwin carried on the work which was to revolutionize our understanding of how life has developed.

was clearly, that on the whole the best-fitted live. From the effects of disease the most healthy escaped; from enemies, the strongest, the swiftest, or the most cunning; from famine, the best hunters or those with the best digestion; and so on.'

Then, suddenly, the idea of the survival of the fittest flashed upon him. Over the next few evenings he wrote it all out carefully in order to send it to Darwin by the next post, which was to leave in a day or two.

Wallace had once met Darwin in the British Museum, and had a general idea that he was interested in varieties in species. Accordingly he had corresponded with Charles in 1857, who realized from his letters and from a paper published in 1855 that Wallace was developing similar ideas. No doubt deliberately, Darwin casually mentioned that he had been working on the species question for some twenty years and that he hoped to publish within a couple of years. But now Wallace had written and his final conclusion, based on a similar understanding of Malthus, was a shattering blow. What was he to do? He could hardly suppress Wallace's paper,

Alfred Wallace sent his paper on variations in species to Charles Darwin for advice on publication. So great were the similarities between the two men's ideas that Darwin was placed in a quandary.

nor could he submit his own. In desperation he quizzed his old friend Lyell on whether or not he could publish his own work:

> 'Wallace says nothing about publication. But as I had not intended to publish any sketch, can I do so honourably because Wallace has sent me an outline of his doctrine? I would far rather burn my whole book, than that he or any other man should think that I had behaved in a paltry spirit.'

At that moment further tragedy struck the Darwin household in the form of scarlet fever. Within days his infant son, Charles Waring, was dead. Darwin was prostrate, and placed the publication matter into the hands of Lyell and Hooker. The solution his friends devised was to read before the Linnaean Society both Wallace's manuscript and an extract of Darwin's paper of 1844. A letter Darwin had written to the American scientist Asa Gray in 1857 was also included to bring the paper up to date. The intended book would have to be forgotten. All was now set to publish the revolutionary new theory.

In the event the meeting of 1 July 1858 was hardly momentous. The papers generated interest and were noted by Richard Owen in his autumn address to the British Association. But when the Society's president Thomas Bell reviewed the year in his annual report he concluded:

> 'The year which has passed . . . has not, indeed, been marked by any of those striking discoveries which at once revolutionize, so to speak, the department of science on which they bear.'

Wallace was content to assume a secondary role. Though, technically, he might have

While Charles Darwin gradually lost his faith in God, his wife Emma remained a pious and devoted believer.

The results of Darwin's life work have placed him in the very front rank of those who have changed our way of understanding the world.

ORIGIN OF SPECIES

In the fourteen chapters of *Origin*, Darwin both explains his theory and counters objections which he knew other naturalists would raise.

The first four chapters contain the core of the argument, opening with a discussion of the way animals of the same species vary and how these variations are passed down from one generation to another. He claimed that the structural differences between varieties within a species and those between species were only a matter of magnitude — given sufficient variations a new species would be born. This cut across the widely held view that somehow each species was different in essence.

Darwin could not demonstrate from nature that accumulated variations led to changes in 'species', so he turned to artificial selection in pigeon-breeding as a base for his arguments. By careful selection and re-breeding, pigeon-fanciers had been able to produce certain characteristics within their birds. By analogy, therefore, any selective force in the wild produced a similar effect, highlighting some variations and suppressing others.

Chapter 3 outlined the struggle for survival in nature as the source of this selective force. Of course, this 'struggle' need not always be a literal 'tooth and claw' fight for supremacy (such as lions fighting over the corpse of a hyena), but simply competition for light among plants, or differing abilities in rearing offspring among animals.

In the following chapter Darwin came to the centre of his theory: natural selection was the mechanism for evolutionary change.

'Can it, then, be thought improbable, seeing that variations useful to man have undoubtedly occurred, that other variations useful in some way to each being in the great and complex battle of life, should sometimes occur in the course of thousands of generations? If such do occur, can we doubt (remembering that many more individuals are born than can possibly survive) that individuals having any advantage, however slight, over the others, would have the best chance of surviving and of procreating their kind? On the other hand, we may feel sure that any variation in the least degree injurious would be rigidly destroyed. This preservation of favourable variations and the rejection

Darwin's seminal 'Origin of Species' contains detailed argument based on an accumulation of biological evidence.

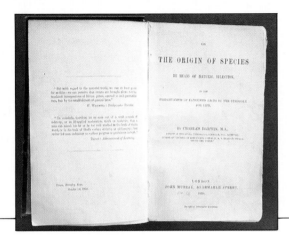

of injurious variations, I call natural selection.'

Darwin believed that such differences between competing animals would be enhanced by the tendency of competitors to seek out different ecological niches, and by gradual geological changes which separate different varieties. But Darwin had no clear understanding of how the variations (whether useful or injurious) were produced, and Chapter 5, which attempted to deal with 'Laws of Variation', was the least successful part of the book. Darwin simply admitted his ignorance.

In Chapter 6 he turned to forestall would-be criticisms. (By the final edition of 1872 this had been expanded to fill two chapters.) Darwin was troubled by the apparent absence of intermediate forms, structures that fitted halfway between two well-known species. It was also difficult to see how natural selection could account for the intricacies of the human eye or how an aquatic animal could *gradually* adapt to live on dry land.

In Chapters 7 and 8 Darwin discussed 'Instinct' and 'Hybridism' respectively. His aim was to discount hybrids as a mechanism for evolution, and to demonstrate how something as complex as instinct may have naturally evolved through selective pressure.

He then turned his attention to geology, pointing out that the fossil record was far from complete. His critics (both then and now) have been quick to point out the lack of an obviously graded fossil sequence in the geological column. Organic forms suddenly appear and disappear. Over time there may be increasing complexity, but from an evolutionary point of view the fossil record is very jerky. Darwin highlighted two problems. The first was the sudden appearance of whole organic groups. Sedgwick and Agassiz regarded this, Darwin acknowledged, as 'a fatal objection to the belief in the transmutation of species'. But Darwin reminded his readers of the chance nature of fossilization, and argued that since the fossil records were far from complete their testimony was correspondingly poor. It was like reading a book with most of the pages torn out.

The second problem was 'much grander'. Why does life abruptly appear in a fairly advanced form? Why do we not find fossilized remains of the swarm of simpler organisms from which everything (supposedly) descended? Candidly Darwin wrote,

'To the question why we do not find records of these vast primordial periods, I can give no satisfactory answer . . . The case at present must remain inexplicable, and may be truly urged as a valid argument about the views here entertained.'

He simply held out hope that further study might answer the difficulty. But Darwin was not going to abandon his new theory simply because one piece of the jigsaw was missing. He had convincing arguments from the geographical distribution of organisms, and his theory could account for this distribution where other theories were at a complete loss. Apart from the fossils everything fitted together so well. And he calculated there was time enough for evolution to have occurred.

In the following chapter he estimated that the strata in Kent were some 300 million years old. This was but a moment in geological time, and Darwin needed long periods of time for his small variations to accumulate into new species. But it was an erroneous calculation he was later to regret. When later estimates, most notably by the physicist Lord Kelvin, failed to support his time-scale, it was the most fundamental blow to his theory. Physics, the king of sciences, could not support its fledgling sister.

Chapters 11 and 12 examined the geographical distribution of organisms. Darwin explained how gradual divergence and migrations would

Cartoonists soon began to see the comic potential of ideas that human beings developed from monkeys, as seen in this shock similarity between a Victorian gentleman and a gorilla.

account for the distribution of animal types around the globe. Such a naturalistic explanation was far preferable, he believed, to those who believed God made multiple creations. How else, he quizzed, could you account for the widely different organisms on either side of geographical barriers such as mountain ranges or deserts? Had God populated one side of the mountain with one set of animals, and the other side with another? Take, for example, the isthmus of Panama. Only the narrow (but impassable) isthmus separated South and Central America, yet 'no two marine faunas are more distinct, with hardly a fish, shell, or crab in common.'

In the penultimate chapter, Chapter 13, Darwin took up the subject of taxonomy and classification. A constant theme of his book was an attack on 'essentialism', the theory that species are somehow eternally distinct from each other. Darwin sought the key to classification in evolutionary

descent, and not in lists of characteristics which distinguished one animal group from another.

In his last chapter Darwin summed up, ending with the frequently quoted words:

'It is interesting to contemplate a tangled bank, clothed with many plants of many kinds, with birds singing on the bushes, with various insects flitting about, and with worms crawling through the damp earth, and to reflect that these elaborately constructed forms, so different from each other, and dependent upon each other in so complex a manner, have all been produced by laws acting around us . . . From the war of nature, from famine and death, the most exalted object which we are capable of conceiving, namely, the production of the higher animals, directly follows. There is grandeur in this view of life, with its several powers, having been originally breathed by the Creator into a few forms or into one; and that, whilst this planet has gone cycling on according to the fixed law of gravity, from so simple a beginning endless forms most beautiful and most wonderful have been and are being evolved.'

Thomas Huxley became known as 'Darwin's Bulldog'. He was also an articulate and committed exponent of the importance of science.

gained priority, his theorizing was the result of one hectic week, not twenty hard-working years. Without Darwin's mass of supporting evidence the 'survival of the fittest' idea would have been dismissed with the same contempt as Chambers' belief that mankind was descended from frogs. The immediate need for Darwin was to publish his supporting evidence. The scientific community could not now wait for his 'big book'; something much smaller (and more readable) was called for.

Accordingly Darwin set to work. The publisher of his *Beagle* account, John Murray, offered to publish the completed work, objecting only to Darwin's inclusion in the title of 'An Abstract of an Essay . . . ' To us, today, a book of fourteen chapters and 155,000 words is hardly an 'abstract', yet this is how Darwin

viewed his manuscript. He feared his argument was inadequately supported, and was surprised when Murray suggested a first edition of 1,250 copies. 'I hope he will not lose,' he wrote to Lyell.

On the Origin of Species by means of Natural Selection, or the Preservation of Favoured Races in the Struggle for Life was published on 24 November 1859, priced fifteen shillings. That the first edition was sold out within the day is true, though the impression given of an excited public queuing in bookshops is false. The first edition was completely taken up by booksellers, who perhaps anticipated the controversy rather than responded to it. The theory was the same as that outlined in 1842 and 1844, though the material was better ordered and replete with examples. In concluding chapters he anticipated objections against his theory and attempted to answer them. Sensitive areas, such as the evolution of humanity, he avoided altogether, preferring to offer hints such as 'light will be thrown on the origin of man and his history', than to revealing further facts still hidden in his notebooks.

Charles sent complimentary copies to his friends. Wallace also received a letter in which Darwin pointed out, 'Remember it is only an abstract, and very much condensed. God knows what the public will think. No one has read it, except Lyell.'

Darwin was anxious to convert the nimble-minded Thomas Huxley. He need not have feared. Writing to Darwin on the eve of publication Huxley vowed to go to the stake in support of parts of the new book. As we shall see, his commitment was more to Darwin's method of doing science, than to his startling conclusions over which his young colleague still had reservations. Nevertheless, Huxley recalls that having mastered the argument of *Origin* he lay the book down reflecting, 'How extremely stupid not to have thought of that!'

5
Understanding the Biology: NATURAL SELECTION

The May 1861 edition of the British magazine *Punch* contained a satire on Darwin's new theory. Under a drawing of an ape carrying a placard with the message 'Am I a man and a brother?', the poem surveyed the scientific ferment aroused by *Origin of Species*. It was signed 'Gorilla, Zoological Gardens'.

Am I satyr or man?
Pray tell me who can,
And settle my place in the scale.
A man in ape's shape,
An anthropoid ape,
Or monkey deprived of his tail?

The *Vestiges* taught,
That all come from naught
By 'development', so called, 'progressive';
That insects and worms
Assume higher forms
By modification excessive.

Then DARWIN set forth,
In a book of much worth,
The importance of 'Nature's selection';
How the struggle for life
Is a laudable strife,
And results in 'specific distinction'.

Let pigeons and doves
Select their own loves,
And grant them a million of ages,
Then doubtless you'll find
They've altered their kind,
And changed into prophets and sages.

The poem went on to mention the studies of Horner, Pengelly and Prestwich which threw doubt on any literal reading of the early chapters of Genesis. It finished by parodying the debate between Thomas Huxley and Richard Owen over whether or not the human skull was significantly different from an ape's.

Darwin's book unleashed a flurry of scientific argument. We have seen how he assembled the pieces of the jigsaw representing a naturalistic explanation of origins. Some of those pieces had already been recognized by other scientists and it was Darwin's genius to link them together. But the resulting picture was not conclusive. Darwin may have delayed publication in order to amass detailed evidence of his theory, yet for many that evidence was still not enough. His book also raised religious and moral questions, but the theory was first and foremost a piece of science and needed first to be judged in that light.

When, years later, Huxley added a chapter to a biography of Darwin, he recalled how the new evolutionary science was first received:

'There is not the slightest doubt that, if a general council of the church scientific had been held at that time, we should have been condemned by an overwhelming majority.'

The crisis in science

At the outset most of the scientists were not on Darwin's side. What, then, were their objections?

☐ *First, there was Darwin's way of doing science.* In the nineteenth century the agreed quest of science was for certainty based upon facts. Now 'natural' selection (some said) was only a guess, or hypothesis, as to how nature worked, and as such was only valid as a tool for further research. A full-blown scientific theory had to be demonstrated by certain proof, and this Darwin could not do. *Origin* was bulging with facts and observations, but the theory based on them was not a law readily observable in the wild. True, the struggle for survival was clearly evident. But had anyone ever watched as a new species was formed? Darwin's theory was only an understanding of nature that could group together observations and explain how they fitted together. It seemed reasonable, given the observed variations in nature and the struggle to survive, to suppose that species evolved according to natural selection. It was just like pigeon-breeders using artificial selection to improve their stock. But this was supposition and analogy, not rigorous proof. And some of the scientists objected.

DARWIN'S SCIENTIFIC METHOD

'But I believe in nat. selection, not because I can prove in any single case that it has changed one species into another, but because it groups and explains well (as it seems to me) a host of facts in classification, embryology, morphology, rudimentary organs, geological succession and distribution . . . ' Letter from Charles Darwin

Darwin believed that natural selection was probable because of the power artificial selection had to change animal forms. Within nature equivalent elements were present, for the ruthless pigeon-breeder was replaced by an equally ruthless struggle to survive. When he began to look at species and animal distribution patterns with natural selection in mind, everything seemed to 'fit'. It was like a crossword: given a few words, the others fitted around them. Of course some elements of nature's crossword still eluded him (such as how variations occur), and for others the clues were half-obliterated (such as in the partial fossil record). But Darwin felt justified in working from the many 'words' he did have to complete the crossword.

Others were not so sure. They were suspicious of an approach that jumped too hastily to conclusions. On the crossword analogy they preferred to wait until all the words had been worked out not by deduction but from the clues. It was invalid, they said, to write in new words simply because the surrounding letters suggested the solution. Darwin must wait until he had observed at least one species of animal change into another, rather than assume that animals evolved because known facts of nature implied as much. To use another picture, it was a bit like catching a thief. Some would only be convinced if he was caught in the act, while Darwin was prepared to accept the circumstantial evidence of footprints, timing and motive.

Darwin's method of working has since been vindicated. Most scientific theories are attempts to link isolated observations into a structure which integrates them together. Certain observed facts suggest a theory which is then tested out to see if it explains other facts. The approach is circular, or rather spiral: facts lead to a theory, which is checked against further facts which subsequently modify and improve the original theory. But almost inevitably there are certain phenomena which do not fit. Recall Lord Kelvin's calculation of the age of the earth. This challenged Darwin's theory because it did not allow sufficient time for evolution to occur. And here Darwin was prepared to admit that some aspects of the fossil record appeared to count against his theory.

In such a position the scientist is left with a choice. Either he allows the unexplained observations to overthrow his theory, or he remains puzzled by them yet maintains his ideas — because they explain so much else. Eventually the criticisms may so mount up that the theory has to be abandoned and replaced with another. (Such was the case with nineteenth-century physics which was eventually replaced by Einstein's theories at the start of this century.) However, until the contrary evidence is overwhelming an inbuilt conservatism hangs on to the old ideas, and seeks to explain the anomalies as best it can.

□ *Second, it was questionable whether variations could accumulate as Darwin said.* Even within artificial breeding, the foundation of Darwin's arguments, there was always a limit to the degree of change possible. A pigeon might be bred to enhance the size of its beak, but continuing breeding soon ran up against a limit of change. And this fell far short of the changes necessary to move from one species into another. The beak might become longer, shorter, curved or straight — but it was still a beak and the creature still a bird. If producing new species was impossible in artificial selection, why should it occur in the wild under so-called natural selection? Furthermore, artificial breeding generated its changes by carefully mating like with like. Even if useful variations occurred in the wild, would they not be gradually whittled away by constant interbreeding with animals without the variation? This was the criticism of Fleeming Jenkin, and it caused Darwin considerable consternation.

Darwin found it difficult to explain how useful variations accrued since he understood inheritance in terms of blending the characteristics of both parents. When children were born of white and black parents was their skin not an intermediate shade of brown? Using the analogy of paint pots it is clear that if, in a dozen pots of white paint, one pot is found to be more magnolia in shade, then subsequent mixing of the pots (interbreeding, if you like) will only weaken the magnolia effect, not enhance it. Darwin replied by suggesting that useful variations might be isolated geographically (that is, all the magnolia paint pots might be set to one side) and that the variations might be fairly common across a large set of organisms. If this were so then it was possible for like to breed with like and for the variation to become well established.

□ *Third, many doubted whether the earth was old enough for natural selection to have achieved such diverse and complex animal forms.* Darwin had suggested an immense age for the earth, judged

William Thomson, the first Baron Kelvin, was a distinguished physicist. His calculations for the age of the earth produced an age too short for Darwin's theories, and this was a damaging blow. Only later did Kelvin's figures prove wrong.

on the speed of rock erosion in his native Kent. The physicist Lord Kelvin, however, derived his calculations from the rate of cooling of the earth. His answer for the age of the earth was different, and far smaller: a mere 100 million years, and insufficient time for evolution to have occurred. It was a serious blow to Darwin's theory. He simply hoped that something would 'turn up' to nullify Kelvin's calculations. The later discovery of radioactivity and of heat-generating radioactive minerals within the earth did indeed dismiss Kelvin's figures, based as they were on the simple cooling of an inert sphere. But they came too late to save Darwin himself from embarrassment.

□ *Fourth, geologists questioned whether the fossil evidence really supported a picture of gradual evolution.* 'We defy anyone, from Mr Darwin downwards,' challenged the *Family Herald*, 'to show us the link between the fish and the man. Let them catch a mermaid, and they will find the missing link.' For many, the geological record spoke of sudden creations and a relatively short time-scale for the earth's formation.

Darwin was more successful countering such criticisms. The geological record was, he claimed, an uncertain witness due to the difficulties of fossil formation. It was like reading a book with most of the pages removed. The story line was uncertain. Darwin confidently expected that further geological research would vindicate his position and (literally) unearth fossils of animals that in form stood halfway between well-known species. The discovery, two years after the publication of *Origin*, of the fossil of *Archaeopteryx* — argued to be a cross between a reptile and a bird — seemed to offer some weight to Darwin's views.

☐ *Fifth, older scientists felt threatened by the new doctrines.* To a degree the arguments of

the scientists related to defending, or enlarging, their own prestige rather than following the noble way of truth. In France the national pride in Cuvier and Lamarck made ready acceptance of a theory by an Englishman very difficult. On the publication of *Origin* not a single leading French biologist came out in support, and full acceptance of natural selection only came in the 1940s. Conversely, as we have seen, many in England were prejudiced against the earlier views of Lamarck precisely because he was French and his ideas smacked of revolution.

More immediate than these social fears was the individual scientist's own position in society.

ARCHAEOPTERYX: THE WINGED REPTILE?

The story of the discovery of *Archaeopteryx* is a tale in itself. The first hint of a species midway between the reptiles and the birds came in 1860 when the imprint of a feather was found in a Bavarian quarry. The imprint revealed little, except that it could be dated long before the agreed 'creation' of birds. Hermann von Meyer called the bird, on the basis of this single feather, *Archaeopteryx lithographica*, which may be freely translated 'Ancient feather from the lithographic stone'.

A year later there was a further find. From the same quarry workmen brought up a slab which showed a nearly complete skeleton, minus the head, of the ancient bird. An opportunist collector, Dr Haberlein, quickly bought the slab, hoping to make money in reselling it to interested palaeontologists. His asking price was too high, and frustrated

scientists were only allowed a brief peep at the interesting specimen. In the end, and much to Dr Haberlein's chagrin, the curator of the Munich fossil collections made a number of these brief viewings and published from memory a drawing of the bird. The doctor was dismayed and eventually sold the fossil to Richard Owen. And so the argument as to whether it was bird or reptile shifted to England.

Some sixteen years later a more complete fossil was unearthed. Again Haberlein obtained it and a similar battle over selling ensued. The German industrialist Werner von Siemens beat Haberlein down to 20,000 marks and passed it on for the same sum to the Mineralogical Museum of Berlin University.

Photographs of the original fossil now appear in most books on

The fossilized remains of Archaeopteryx, which was believed to be a missing link between reptiles and birds.

evolution. The argument as to whether or not it represents a genuine transitional form between bird and reptile, and hence provides evidence of evolutionary theory, continues to this day. In 1985 Fred Hoyle and Chandra Wickramasinghe claimed the fossil was a blatant forgery. Haberlein, they said, had assembled some dinosaur bones and added feather marks by pressing real feathers on to a paste of powdered limestone. The present owners of the fossil, the British Museum of Natural History, remain unswayed, and the museum bookshop sells a booklet to convince any sceptic.

It needed to be guarded. In the heat of the nineteenth-century debate, scientists who had laboured all their lives in the establishment of an alternative view would not on the reading of a single book abandon deeply held and cherished beliefs. They were committed to supporting the fixity of species.

Quite properly the new ideas needed to stand the test of time. Perhaps less properly, the up-and-coming generation of scientists (Darwin, Huxley, Tyndall) were anxious to establish their reputations, just as the older scientists, such as Owen, were concerned to defend theirs. Sometimes personal rivalries played their part. Richard Owen was one of the few who had spoken warmly of *Vestiges*. He might have been more favourable towards *Origin* if he had not been goaded over other matters by the younger Huxley. Huxley supported Darwin, and so Owen was forced to take sides, perhaps against his better judgment. Coupled as this was with an increasing tendency to 'professionalize' science, to rid it of its clerical and aristocratic background, the scene was set for a cultural conflict masquerading as argument over scientific facts. The question was not only 'who is right — Huxley or Owen?', but also who would dominate science and whose name would endure.

In the 1860s there was a crisis of faith. But it was as much within the citadel of science as among the camps of the church. The general public was guided in its response by its scientific mentors. Alongside the scientific criticisms there were objections to *Origin* on religious and moral grounds and we will examine these in the next chapter. Darwin's book implied that human beings were just advanced apes and design no more than fortuitous chance. This belittled humanity as made in the divine image and questioned the hand of God in creation. Yet it is a caricature to believe that the conflict was centred in an ignorant and bellicose church trying to stamp out the brightly burning flame of science. Yes, priests declared Darwin a heretic, while Darwinian disciples such as Huxley preached the new doctrine of

The Royal Institution, the lecture-room of which is seen here, was a focal point of the nineteenth-century scientific enterprise.

all-pervading naturalism. But there were many reputable scientists who questioned Darwin's conclusions, just as there were churchmen who accepted them.

But now we must take a break from the Victorian debates in order to understand Darwin's theory more fully. And rather than review the theory again later in the book, we will use here some examples from our present-day understanding of natural selection. Darwin's theory revolved around a number of key ideas:

☐ *The tendency of organisms to multiply to fill their surroundings;*
☐ *The competition within nature for available resources;*
☐ *The variability of individuals within a population, leading to natural selection;*
☐ *The gradual adaptation of organisms to their environment;*
☐ *The production of apparent design in nature.*

EXPONENTIAL GROWTH OF POPULATIONS

In mathematics a series of numbers which show a regular increase is called a 'progression'. The type of progression with which we are most at home is called the Arithmetic Progression, in which the numbers increase by a constant amount each time. For example:

2 4 6 8 10 12 14

But there is another type of progression, very common in nature, in which the numbers increase by a constant multiple. This progression is called a Geometric Progression and would look like this:

2 4 8 16 32 64 128

As you can see, the numbers increase much more rapidly in the geometric progression than they do in the arithmetic progression.

You may have noticed in the main text a story of the office boy's wages. That story gives an example of both types of progression: the days increase arithmetically but the wage increases geometrically. For those who would like to follow the sum through, the table below shows the boy's daily wage over a 28-day period.

Days	Wage in Pence
1	1
2	2
3	4
4	8
5	16
6	32
7	64
8	128
9	256
10	512
11	1024
12	2048
13	4096
14	8192
15	16384
16	32768
17	65536
18	131072
19	262144
20	524288
21	1048576
22	2097152
23	4194304
24	8388608
25	16777216
26	33554432
27	67108864
28	134217728
Total	268435455 pence

Nature's multiplication tables

Darwin received his inspiration from Malthus. Looking around at nature it was obvious that there is severe competition for available resources. Trees in a forest struggle to reach upwards for the light, while in a severe winter birds die for lack of food. Malthus spoke of food supplies only increasing arithmetically, while populations increase geometrically. What did he mean?

An office boy, seeking employment, said to his prospective employer, 'Start off by paying me a penny a day in wages. If you are satisfied with my work at the end of each day then double my wages for the next.' Believing he had a bargain the employer took the young man on. His work was exemplary and at the end of each day his wages were doubled. But at the end of a month the employer was faced with a bill for £2,684,354.55 (see the adjacent feature article, *Exponential Growth of Populations*). And all this was for the office boy alone! This is an example of what is called geometric or exponential growth and it is this kind of growth which is found in populations of plants and animals. If it continues unchecked such growth reaches staggering proportions in a very short time. As an example consider the common bacterium *Escherichia coli* which under normal conditions will divide every twenty minutes. Over a two-day period this means 144 consecutive divisions. Each bacterium weighs only one million millionth of a gram, but within this two-day period the total weight of bacteria produced will exceed the weight of the earth! Clearly this is nonsense, and in the real world something prevents the total population from achieving anything like its theoretical potential.

In fact, population numbers rise rapidly to a maximum and are then maintained, within quite fine limits, at a value known as the carrying capacity for that environment. In the case of bacteria two things limit their numbers: shortage of food and pollution of their environment with excreted wastes. In the case of plants, we see similar limits to population size set by the availability of water, light or sites in which to grow. Similar constraints exist for animals, but we also find 'behavioural' limits. For example, a number of bird species will not attempt to breed unless they have a safe site in which to nest. But there is something beyond this. Wherever there is a resource which is in limited supply the consumers of that resource will be in competition with one another. And once one individual has consumed that resource it is no longer available for others. This may seem self-evident but the consequences of such competition for a resource, be it food, water, or a place to nest, can have profound effects on the individuals engaged in the struggle.

In studying the structures of communities from very different parts of the world, biologists have had to face the fact that the species in each community are very different. However, within each community there are certain roles which must be filled if the community is to function properly, and these roles are termed 'niches'. In all communities we find primary producers, primary consumers, secondary consumers, and the 'detritivores' (such as scavengers) which remove the waste. Thus the biologist can identify the role an organism plays within a community rather than identifying each individual species by name. It is like saying 'There is the butcher' rather than 'There is Mr Jones'.

The niche concept has been extended beyond the idea of a food-related role in the community to encompass all the factors which control an organism's life. The temperature range in which a species can survive, or the humidity level a species needs, can also be considered as part of the organism's niche. Once these environmental features are taken into the picture, along with the biological components of the community, then scientists term the whole thing an ecosystem.

Only a limited number of individuals, then, can survive in the same environment. In the same way only a limited number of species can be 'packed' into a given ecosystem. Their niches must be separated from each other either by a physical difference, or by a behavioural difference, or by a time difference. To give an example, four species of bee can be found in the same region apparently living on the same food. But closer inspection reveals how one species has a much longer tongue than the other three and can reach parts of the flowers that the other bees cannot reach. Another bee is never found far away from woodland while a third is always found in the open fields. And the final separation is on the basis of time: the last of our four bees is active late in the season unlike the others which swarm earlier in the year. In this way the individual niche of each species remains separate and the resources are partitioned between them.

But where the niches of two species do overlap there will inevitably be competition between them. Equally inevitable, there will be a winner and a loser. Naturalists observe that two species cannot coexist on the same limited resource, and the weaker will be eliminated by the stronger. One species is driven from the overlapping part of its niche into the remaining area, to what is called its 'refuge niche'. In these ways populations come to occupy only a certain part of their total environment. Their position is limited by the physical and chemical features of their ecosystem and they are pressed from all sides by the other populations in that community waiting to take over and occupy their space.

The struggle to survive

The most obvious struggle within nature is that between a predator and its prey: the members of one population eat the members of another population for food!

The relative size of the predator and its prey is important in determining the type of hunting strategy adopted. If the predator is roughly the same size as the prey then a one-to-one chase and capture technique may be used. When the prey is larger than the predator then the predators may hunt in packs. The herd of animals which a lion is hunting may be attacked from one side by a single lioness and set running towards a line of waiting lions. These lions isolate a single individual from the herd and overwhelm it by strength of numbers. After the kill the prey is shared between the hunters so all benefit from the hunt.

The exact opposite is also found, where one very large predator feeds on a vast number of much smaller prey filtered from its surroundings. The blue whale, the largest living mammal, consumes shrimps called krill. Biologists estimate that the stomach of one 26-metre individual contains about 5 million krill weighing two tons!

Whichever strategy is adopted many of the features of the predator's body will be geared towards it. The teeth of a dog are modified with large canine teeth for holding on to struggling prey, and the molars and premolars are designed to cut flesh rather like kitchen scissors. The rabbit, which feeds on vegetation, has large front teeth which will cut through leaves. The molars

COMMUNICATION AND SOCIAL INTEGRATION

Animals may be at an advantage if they are able to reduce the level of competition between members of their own population. We can sometimes find very sophisticated kinds of social integration where co-operation between individuals in a population requires some means of communication between them.

The form of communication used often depends on the type of environment. In dense jungle, where visual range is limited, an animal such as the howler monkey uses its voice for communication. The hooting of an owl at night establishes the limits of its own particular territory in relation to its neighbours.

When the environment is more open a visual means of communication may be employed, and the flocking behaviour of many birds or the tendency of some fish to form schools are examples of such visual communication. This communication can be quite sophisticated even in the lower animals. A bee can communicate the location of food to the other bees in the hive by a dance it performs on the surface of the honeycomb.

A further type of communication which is found throughout the animal kingdom uses chemical signals, often called pheromones. Leaf-cutting ants in the tropics will carry small pieces of leaf back to the nest, leaving as they do so a trail of scent. Fellow-ants leaving the nest on a foraging trip will follow this chemical trail back to the source of food and on the return journey they too leave a scent trail for others to follow. Once the food supply is gone the ants stop laying a trail and the volatile marker-scent soon evaporates so no further ants are directed to the now-exhausted food plant.

A clear example of social organization among creatures is the ant. Every aspect of an ant's life is conducted in co-operation with others of its species.

The marking of a territory by chemical scents is very common. Rabbits have a gland under their chin which is used to mark patches of grass, and deer have a gland just under the eye which they rub against tree branches to mark them. Scent-marking is the reason your dog stops at every lamp-post and tree when out for a walk. Chemical communication can even cross the boundaries between two populations. The reproductive cycle of certain fleas is controlled by the hormones in the blood of the rabbits on which they live, ensuring that the young fleas are hatched at the same time that the baby rabbits, on which they will live, are born.

Lions are among the most accomplished predators. The struggle between predator and prey is a clear example of competition between populations within a habitat.

are flat and ridged for grinding through the tough walls of the plant cells to release the cell sap. And, since the rabbit does not chase and catch prey, where the dog has its canine teeth the rabbit simply has a gap.

A predator also needs to sense direction, and to judge distances accurately. The owl has eyes set in the front of its face, providing it with good binocular vision for hunting. The rabbit's eyes are set in the side of its head providing it with good all-round vision for detecting would-be predators. In the low visibility of a muddy river the electric eel can detect disturbances in its electric field caused by the small fish on which it feeds. The spider can locate the part of its web in which an insect is trapped by plucking the lines of its web and finding which ones are dampened by the body of its prey.

The other side of the coin is the adaptation found in species to try to prevent them from being captured. There are three options for a prey animal in the presence of a predator: fight, flee or hide. Fighting is not a very common practice, for the majority of animals which graze on plants are ill-equipped for it. But the few which do have horns, especially when allied to a large and powerful body, may be successful.

However, as Mark Twain observed, 'Presence of mind in an emergency is admirable, absence of body is preferable.' Most animals choose to flee rather than fight. To this end the legs of a creature such as the horse are long, slender and very powerful, and its lungs can take in enough oxygen to allow it to run at speed for some considerable time. It can outrun its predators.

Interesting as these adaptations may be, it is in the strategy of hiding that some of the most ingenious adaptations are to be found. The pale-coloured belly of many animals is in sharp

MIMICRY

Some animals bluff their way out of trouble — they mimic other species which are known to be distasteful or toxic. It is not uncommon to find that toxic animals have striking colour markings on their bodies; the black and yellow bands on a wasp are a good example. Once a predator has had a few unpleasant encounters with these animals, it comes to recognize the distinctive pattern and to avoid them where possible. It is quite literally a case of 'once bitten, twice shy'!

Where such distinctively marked toxic animals are found it is common to find other animals which mimic them. Wherever you find the wasp

Camouflage is a common strategy for animals to adopt in their battle to survive.

you will also find the hoverfly, a harmless insect but one which has also adopted the distinctive black and yellow stripes. At a casual glance most people would think that the hoverfly

was a wasp — and fortunately for the fly so do potential predators. Provided enough of the truly toxic wasps exist in the area to keep reminding predators of the unpleasant consequences of eating them, the mimics are protected from attack.

contrast to their dark-coloured backs, a characteristic called counter-shading. When sunlight falls on the top of the animal its back is in bright light but its underside is in shade. But with a counter-shade pattern this produces an even light intensity from top to bottom. Most animals are colour-blind so the difference in colour is less important than the fact that the even light intensity conceals the rounded form of the animal's body. The shading makes it look much flatter and, hopefully, confuses a predator.

The stripes on a tiger and the spots on a leopard are both examples of what are called disruptive patterns. Not only can they resemble the dappled patterns of light and shade found in the animal's natural habitat but they also disguise its outline by blurring the edges of its body. Modern armies use the same principle in the combat dress of their soldiers.

Some animals resemble their background so well that even when you know they are there it can be quite difficult to spot them. If you walk across a moor a grouse, motionless

in the heather, is very difficult to detect and it can startle you when it finally flies up almost from your feet. Other animals are even better at disguise: they can change until they resemble the background they are on. Some flat fish can change the colour and pattern on their backs, over a period of several minutes, to resemble either the gravel or the plain sandy bottom on which they are resting.

Still other animals resemble inanimate objects: some caterpillars can look like bird droppings or twigs, while butterflies can look like dead leaves as they rest on a twig with their wings closed. As long as they remain still their disguise is most effective. Should they be detected and attacked, animals such as the butterfly may try a bluff technique. It is common to find on the lower wings of a butterfly large 'eye spots' which would normally be hidden. But in the event of an attack the victim suddenly reveals them, and the effect on the predator of suddenly being confronted by what appears to be the eyes of a very large animal is quite disconcerting. Even a momentary confusion in the attacker may give the prey time to escape.

Though a plant is far more tolerant of being eaten than is an animal, even they have defence mechanisms: the thorns of a rose bush or a bramble, for example. But a much more subtle mechanism is used by the oak tree which has tannins in its leaves. These are complex chemicals which can denature proteins and render enzymes inactive. A caterpillar eating oak leaves will have some of its digestive enzymes put out of action. Though it eats the food, it gains no nutrients from it and so will fail to grow to maturity and reproduce. This does nothing to stop the immediate problem of leaves being eaten but it cuts down the number of caterpillars feeding on the tree the following year!

Not all populations are kept in check by predators, however. If you look at the records of the Hudson Bay Trading Company over a 100-year period you will notice a cycle in the numbers of trapped snowshoe hares and lynx. The explanation put forward is that any increase in the number of hares leads to an explosion in the lynx population, which feeds on the hares. The more hares there are the more lynx that can be supported. But then the numbers of lynx begin to outstrip the available hares and the subsequent over-feeding causes a decline in the snowshoe hares. The lynx population is then starved down to low numbers. With the relaxing of the level of predation the population of hares once again increases, which in turn supports a larger lynx population. And so the story continues, and it is a standard account of predation maintaining a balance within an ecosystem.

There is one problem: the population of snowshoe hares on the Anticosti Islands in the Gulf of St Lawrence also follows a regular pattern like those on the mainland. But there are no lynx on the Anticosti Islands and, in fact, there are no large predators at all! So why does the snowshoe hare population fluctuate here as well?

The answer comes from a study of voles near a lake in Wales. Here the voles have abundant food and no predators but they still have a population size which varies over a four-year cycle. The study showed that whenever the numbers in the population began to drop most of the animals were suffering from a kind of 'stress syndrome'. This lowered their birth-rate and increased their susceptibility to disease. But these individuals were genetically different from those in the population when the total population was small.

At the times of maximum numbers the individuals had a very high birth-rate, but they could not cope with the stress of crowded conditions, the fight to establish a piece of territory, and the struggle to find food. The high birth-rate and

The wings of this butterfly carry false eyespots, which can be suddenly revealed to confuse an attacker.

stress sensitivity were genetically linked; one always came with the other. And this meant that the reverse was also true: the ones which were able to cope with stress had a very low birth-rate.

The net result was that as the high birth-rate individuals died out there was an inevitable reduction in the size of the population as the reproductive rate slowed down. Eventually numbers would reach such a low value that the few surviving high-birth-rate individuals were at an advantage and their numbers increased over the numbers of the low-birth-rate voles. Yet this increase in numbers carried with it the seeds of its own eventual disaster, for once the population size had grown sufficiently large the high-birth-rate individuals were once again unable to cope with the crowding stress and died.

In this way a cycle was maintained in the population, which oscillated between the two different genetic make-ups. It is a state called by population geneticists 'balanced polymorphism'.

We see that when the population density is greater than the carrying capacity of the environment factors within the population tend to reduce the birth-rate and increase the death rate. When the population density is below the carrying capacity the same factors tend to increase the birth-rate and decrease the death rate. Either way the population is stabilized at a density around that of the carrying capacity. In the case of the snowshoe hares and the voles the controlling factors are linked to the stress that individuals encounter in fighting for a territory which in turn is linked to their ability to reproduce.

In northern Canada the snowshoe hare is preyed on by lynx. The populations of hares and lynx go up and down in an intricate balance.

Variations and inheritance

Even in a crowd we can all pick out our mother or father, brother or sister, or our next-door neighbour. Though we are all human beings there are sufficient differences between each of us for us to recognize each other. How? The answer is obvious: with the rare exception of identical twins no two individuals are exactly alike. The same is true of other animals, and we give individual names to our pet dog, cat or tropical fish. There may be a similarity in the overall pattern of the individuals in one species but the fine detail is very diverse. Every human head has two eyes, a nose, a mouth, two ears and hair. But an infinite variety can be woven by changing the colour of the skin or hair, or by altering the distance between the eyes or the width of the mouth.

These physical characteristics are passed on from one generation to the next. A proud grandmother will gaze at her new grandchild and say, 'Doesn't he have mother's nose and father's hair?' She is expressing what is a commonly observed fact, that children bear a resemblance to their parents.

For hundreds of years, possibly thousands of years, farmers have been using these two facts of variation and inheritance in a practical way. By selecting for breeding only the biggest and the best individuals, the farmer can create a herd in which the average size of each individual animal is greater than before. The same is true for crop plants: by sowing only the seeds from the most productive, hardiest and healthiest of plants the farmer can increase the size and productivity of his crop. He draws on the few valuable variations which already exist in his herd or crop and breeds these so that they become the most usual form.

The variability hidden in a population of domestic animals or plants (and revealed by selective breeding over many generations) is not limited simply to size. An enormous range of fancy pigeons has been bred from the single species of stock dove. As Darwin knew, fanciers could rear not only birds with different colouring, but strains which showed quite remarkable changes in form. Some possessed fantails, others boasted neck ruffs.

We all know, then, that small variations exist between the individuals in a population. Given this, it is not unreasonable to assume that some of these differences could affect the chance of any one individual surviving. Or the differences might affect an individual's fertility. Each organism is subject to competition from other organisms, and here we can draw an analogy between the effects of competition in the biological world and the effects of competition in the world of big business. A company which is uncompetitive soon goes bankrupt, and an organism which is uncompetitive soon becomes extinct. This is an analogy which would not have escaped Darwin who lived in the competitive capitalist society of Victorian England. Those individuals best suited to a particular environment have a higher probability of leaving offspring than do the less well-adapted individuals. Of course, within a population of domestic plants or animals the selection of individuals as future parents is done by the farmer. But within a natural population, selection of those individuals which will breed is achieved by the types of environmental pressures we have already described, namely predation and competition.

The environment in any particular place is never constant. But then neither is a population, and as the environment changes a new set of 'best-adapted individuals' is selected from within the population. It is these individuals which make the greatest contribution of offspring to the next generation. Provided the changes in the environment occur on a time-scale comparable to, or longer than, the average time between generations, then adaptive changes can occur. But if the environment changes more quickly than the span of one generation of the species, then individuals have one of two choices. Either they must become acclimatized to the changes — rather like mountaineers who have to rest for several weeks at high altitude to get used to the rarified atmosphere — or they have to move from that environment to one which they can tolerate, just as birds do when they migrate to warmer climates in the winter.

Bacteria and black moths

In some species the life-cycle is so short that successive generations can evolve to become adapted to what we would regard as rapid changes in the environment. Some flies move

Some species of birds are able to tolerate the climates of the different seasons in a habitat. But others have to migrate to warmer parts in winter.

through several generations a year and each successive generation is adapted to the conditions of a particular season! Within a few days bacteria can become adapted to life in an environment containing antibiotics. They evolve into the 'antibiotic-resistant' strains of microbe which present such problems for both patient and doctor.

This last example is very useful for illustrating the whole process of adaptation to the environment. We have seen that within a population there is a hidden variability which can be revealed by selection. Within the population of bacteria there are a few individuals who happen, quite by chance, to be resistant to penicillin. Under normal conditions this resistance is of no advantage to them at all. But when their environment is changed by the presence of penicillin they alone of the bacterial population are selected to survive. Not only are they the only remaining members of their population capable of reproducing, but also have the added advantage that the number of potential competitors has been reduced. Under these conditions a rapid expansion in numbers can occur, swiftly producing a population in which all the individuals are resistant to the antibiotic.

This is a very clear example of adaptation to the environment: the non-resistant individuals are killed by the antibiotic and only the resistant ones survive. But this is not always the case, for there are many situations where the total destruction of one group of individuals is not the important feature. Both forms may survive but their contributions to the next generation will differ. The change required to adapt the population to a new environment is gradually accomplished by altering the relative percentages of the two types of individual in the population.

We can illustrate this type of adaptation by considering some experiments done by H.B.D. Kettlewell in the early 1950s. He was studying a moth called *Biston betularia*, or the Peppered Moth, which can be found in woodlands in Great Britain. The normal wing pattern for this moth is speckled grey-brown, but some melanic

individuals, which are completely black, occur by chance in the population.

Before Kettlewell began his experiments he knew that the offspring of a melanic moth would also be black: the melanic trait was inherited. He also knew that the earliest examples of melanic individuals had been recorded from forests in or near industrial areas and that the highest frequency of melanic forms in a population was currently to be found in such woods. Where there was little or no industry the normal, light form of the moth predominated.

Kettlewell collected and raised more than 3,000 caterpillars to provide adult moths. These he marked, on the underside of their wings, with a small dab of cellulose paint. In his first experiment he released equal numbers of normal and melanic moths in a polluted wood in the industrial midlands of England. After a short time he set about recapturing the moths he had released. He wanted to find out how many had survived. For every 100 moths of the two types released, Kettlewell recovered sixteen of the normal moths but thirty-four of the melanic variety. Twice as many melanic moths had survived as normal ones.

In a similar experiment in an unpolluted forest in the south of England, Kettlewell recaptured thirteen of the normal moths as against six of the melanic form. Here twice the number of normal moths had survived when compared to the melanic ones — the complete opposite of the results from the polluted wood.

In looking for an explanation Kettlewell considered the differences in the two habitats. The tree trunks in the unpolluted woods were covered with a grey-brown blanket of lichens. However, lichens are very sensitive to sulphur dioxide in smoke so the trees in the polluted wood had trunks which were bare of lichens and had become blackened with soot from factory chimneys. Here was the key to the problem. The normal form of the moth was well camouflaged against a background of lichens, but against a black, soot-covered tree trunk it was easily spotted by birds and captured as food. The melanic form, on the other hand, was camouflaged on the polluted tree trunks but readily spotted against a lichen-covered background. Kettlewell went on to make detailed observations from a hide and confirmed that the differences in survival of the two forms of moth were related to their predation by birds.

We can see, then, that if at each generation we have half the numbers of the normal type surviving compared to the melanic form, then over a number of generations the melanic form will come to be the dominant type in the population. Where the environment favours the normal type the opposite occurs and the normal form is dominant. Selective predation can have a big impact on the make-up of a population in nature as well as in the farmyard or pigeon loft!

How many moths can you see? Black peppered moths began to develop alongside speckled ones when smoke pollution produced blacker trees on which they could not be seen by predators.

Natural selection and reproduction

In nature there are two basic approaches to reproduction:

☐ *To produce literally thousands of offspring and accept a very high infant mortality;*

☐ *To produce a few young and look after them until they are big enough to fend for themselves.*

These approaches an ecologist calls reproductive strategies: the herring is an example of the first, and the shark of the second. The herring lays millions of eggs of which only a small fraction survive. But the shark lays only a few eggs and provides each one with a yolk to nourish the growing infant, so that when the baby shark hatches it is already several inches long and has a much better chance of survival. In between these two extremes there are many animals which lay varying numbers of eggs and provide varying degrees of care for the offspring. But there are no fish which combine the extremes of both strategies, laying millions of eggs and providing each with a yolk. The reason is simple economics. The mother has to manufacture both the egg and the yolk from the food she eats, and this physical limitation sets the limit to her reproductive potential.

If natural selection is as potent an evolutionary force as suggested then there should be examples in which the reproductive strategy of different populations of the same species has changed in response to different selective pressures. And there are indeed such examples. Studies show that lizards on the Italian mainland lay a larger number of eggs than lizards on the offshore islands. On the mainland, food and water are in plentiful supply and the size of the lizard population is controlled by predators. These mainland lizards produce between four and seven eggs and so provide a margin of eggs above what is needed to replace the population, for some eggs are always eaten by the predators. Now consider the islands. There are no predators but instead the availability of food and water set the limit on population size. The eggs can therefore be fewer in number, between two and four in a typical clutch, but they need to be larger than those on the mainland. The extra size gives sufficient food to the infants during their vulnerable early days of life.

For some birds the limit to the size of brood they produce is set by their ability to feed the chicks. The great tit will lay between eight and twelve eggs in a clutch depending on the availability of caterpillars at the time the eggs hatch out. The eggs are laid long before the caterpillars are around so the bird 'predicts' the future availability of food. How is this done? Well, the average temperature in March affects the number of caterpillars which will be available later in the year. The great tit has evolved a mechanism which also responds to the March temperature and she lays an appropriate number of eggs. This is not choice on the part of the bird. There is no careful calculation of whether this year will yield a good crop of caterpillars! Its reproductive organs respond automatically to the March temperature to vary the number of eggs produced.

In the case of the Italian lizards the number of eggs produced was set by the hereditary information in the lizard's genes. In the case of the great tit the number of eggs produced is in response to an environmental factor, but the way in which the bird responds to that factor is programmed into her genes.

To change or to stay the same?

Natural selection can operate on the genes present in a population in three basic ways:

□ *Stabilizing selection* acts to maintain a constant genetic make-up in the gene population by eliminating those individuals which are different. For example, in 1899 H.C. Bumpus collected and measured sparrows which had been killed during a violent storm. He found that far more dead birds had either short wings or long wings than average wing length. The storm had a much smaller effect on those individuals which were 'normal', but it eliminated a larger proportion of the two extreme forms of sparrow.

□ *Directional selection* results in a change of the make-up of the genes in one direction in respect to particular characteristics of that species. In a long-running experiment started in 1895 by workers at the University of Illinois, artificial selection was applied to a field of corn to increase the oil content of the grain and reduce the protein content. They continued for over fifty generations to get steady changes

in both these characteristics. They then applied reverse selection and restored both oil and grain content to their original levels, so demonstrating how much variability still remained within the genetic make-up of the corn population.

Under less controlled conditions the widespread use of pesticides after the Second World War has had marked effects on the genetic make-up of some insects. At Clear Lake in America an attempt was made to eradicate the midge population by treating the water with DDD, an insecticide closely related to the better-known DDT. After three treatments over a period of several years the midge population was as large as ever and resistant to the DDD! Unfortunately many of the fish in the lakes and the birds which fed on them were not resistant and died as a result.

Though neither of these examples could be described as 'natural' selection they do illustrate quite graphically how rapidly the effects of selection can occur. Natural selection with a directional effect will tend to occur if a population undergoes a progressive change in its environment. For example, the climate may become colder or drier and organisms will need to adapt to survive. In a piece of genetic detective work, geneticists have shown that the grasses found on the Great Plains of America form a regular series spreading out northwards and westwards from the Mississipi Valley. In each zone the individuals are genetically adapted to conditions of increasing cold and drought.

□ *Disruptive selection* acts to break up a single population into a series of sub-populations when changes occur in a section of their environment. A stream may overflow its bank and flood part of the habitat of a plant. The change may affect only some of the population while the remainder still lives in its normal habitat.

G.L. Stebbins studied one such process in a population of sunflowers in the Sacramento Valley in California. During a period of twelve years the original single population, which was a hybrid between two species, split into two sub-populations separated by a grass patch over fifty metres wide. One population was adapted to dry conditions and resembled one of the original parent species from which the hybrid had been formed; the other sub-population became adapted to wetter conditions and continued to resemble the original hybrid form. Sunflowers are normally cross-pollinated by bees and although this exchange of genetic information undoubtedly took place between two sub-populations the force of the selective pressure was strong enough to maintain them as genetically distinct entities.

Here, in disruptive selection, we have the starting point of speciation — the mechanism by which a new species is formed. If two sub-populations continue to diverge and adapt to different conditions, then the eventual product may be two distinct species which can no longer interbreed.

The blind watchmaker

We human beings tend to think as individuals, and so, quite naturally, when we think about biological situations, we identify with the fate of the individual. We think in terms of a particular great tit — the one which nests in our garden or a nearby hedgerow. But natural selection does not work like this: it operates on the variations between individuals in any population to give direction to the development of that population as a whole. It channels the population into better adaptation to its environment and tends to be rather harsh on the individuals. This produces the best chance for the population to survive.

Put that simply, many people have found it hard to believe that the rich diversity of plant and animal life could have arisen in this way. Archdeacon Paley insisted that the existence of a watch was evidence of the activity of a watchmaker, rather than the consequence of chance, and so inferred that nature is designed by God and is no accident. But this illustration in no way rules out evolutionary development, because Paley's watch is also the result of such a process. It did not spring fully formed into the world; it has a traceable history. Generations

Whether or not biologists believe in a Designer, knowledge of how organisms work does not lessen their sense of awe at the intricacy and beauty of a bird's wing.

of watchmakers selected those basic physical principles which would achieve their aim of making a mechanism to monitor time. They tried and discarded many species of timepiece until they finally refined the mechanisms of which Paley's magnificent watch was one. The watchmakers had selected, from the variety of methods available, those which were best suited to the task of making a small, accurate watch. All the unsuccessful methods — like water clocks, pendulums, or marked candles — have now become extinct. We see them in museums much as we see the skeletons of dinosaurs.

In the biological world Richard Dawkins has likened natural selection to a 'blind watch-maker'. It 'selects' from the variability available in each generation those individuals which are best suited to the task of surviving in that environment. According to Dawkins, the difference between the watch and living organisms is that the watch was designed for a specific purpose. It is there to tell the time. But organisms have only one 'purpose' which is to survive. What they look like and how they function does not matter. The blind watchmaker works by selecting out those individuals which do best at whatever it is they do and develops subsequent generations from them. Those which do badly are discarded.

When we look at a beautiful rose or admire the flight of a swallow, we can hardly help but think of design. Even if, like Dawkins, we reject the idea of a creator God, it seems as if the scent of a flower or the wings of a bird have gradually been perfected towards their final design. God may not have done it, but nature certainly has. The goal of attraction by scent, or movement by flight, was there from the start.

But some biologists would say that in the real world of living organisms there is no such goal. A species continues as a species just so long as there are individuals within it capable of producing the next generation. Of course all species live under certain constraints: the physical and chemical properties of the constituents from which they are made, the cellular nature of life, the aggregation of cells into tissues and the interactions of individuals in populations and communities. But within these constraints the exact form and function used by a population of individuals to produce the next generation is open to a number of options. It is never determined beforehand.

Let us take an example to illustrate this. It has been suggested that evolution is as likely as setting a driverless car going in Paris and for that car to reach Berlin. With no driver the idea is absurd. At the very first bend it would crash and its progress would be measured in metres, not kilometres. But all this assumes that there is a final goal to be reached, namely Berlin. And as we have said, many biologists no longer think in these terms. They say that in the real world you set your driverless car going in Paris and then, provided it is still going in an hour's time, wherever that car is found you call Berlin. When we identify a species we look at where it is now, we do not compare its position relative to some arbitrary target. If the car

crashes before the hour is up it has effectively become 'extinct' in biological terms. This was the deep question raised by Darwin's new theory. The whole matter of design in nature and the existence of a Designer was questioned. Complex organisms could be seen as no more than random movements of molecular structures across the map of life.

Before we leave our car illustration we should perhaps take one final look. We have talked of the car moving in a random way, as though evolution is a random journey. But the route taken by a driverless car will never be completely random: it will be constrained by the hills and valleys of the ground over which it must travel. For its route to be completely at random you would have to start it going in the middle of a vast expanse of perfectly flat concrete — not a very realistic view of the real world. Just as the driverless car will be limited in where it can travel in a landscape of hummocks and hollows, so too the biological forms taken by the individuals in a population are, at a very fundamental level, limited by the interactions which are possible between atoms and molecules. Chemistry writes the rules. These restrictions in turn limit the structure and function of the parts of the cell which determine the joining together of cells into tissues, and so on up the hierarchy of the living world.

Working within the necessity to follow the physical and chemical laws of the world, there are only a limited number of structures possible. But there are alternatives at each step of the way. Which alternative is chosen may happen by chance, so the final outcome is never determined beforehand. Consider the moment when our driverless car reaches a point where two valleys divide off from its route. Which valley will it go down? Perhaps the car will hit a pot-hole or strike the kerb and will be knocked into a particular direction. How the choice is made matters less than what final destination is reached, and depends totally on which valley the car in fact goes down. Had the car taken the alternative valley it would have ended up somewhere completely different. In the biological world the constraints of physics and chemistry may allow evolution to travel only along certain 'routes'. There may be many possibilities at every stage, but there is not an infinite choice.

So the role of chance may be constrained by the physical world. Other scientists go further and see the 'accident' which determines the direction at key points as also designed. Perhaps evolution is not, after all, subject only to blind chance. And our analogy assumes there is no driver whereas some see within nature a guiding force. They accept the role of natural selection but believe that the variations which feed it are determined by a mysterious life force, or by God.

A sufficient theory?

We have surveyed some of the elements of natural selection in action. But these details do not give clear evidence that it is this mechanism, and this mechanism alone, which is responsible for changing one species into another. Indeed, in the examples given, we have not discussed evolution from one animal to another. We can point to small changes as organisms adapt to their environment, and then go on to suggest (as Darwin did) that these small changes can accumulate. Eventually the small changes add up to a significant change, and a new species is born.

But Darwin's critics believed natural selection was only a conservative force. They would not have been surprised to learn that it could account for the changes in the balance between two or more forms, the degree of change we see between the normal moth and its melanic cousin. Natural selection can perhaps account for these population changes but the speckled and black moths are both still moths. In Kettlewell's experiments a moth never turned into a Monarch butterfly, let alone a butterfly into a bird. The long debate over Darwin's theory is whether it is sufficient on its own to explain the diverse organic forms we find on our planet. For if natural selection is not the mechanism, what is?

In *Origin of Species*, Darwin never claimed he had *proved* that species had evolved by means of natural selection. He firmly believed he had discovered the mechanism, but demonstrations akin to those found in the physical sciences were clearly impossible. He highlighted the existence of selective pressures and so moved on to propose that, if small adaptive changes could accumulate, evolution was a logical outcome. The theory was, and is, widely accepted because it explains so much of what we see in nature. Where scientists had been puzzled — by the geographical distribution of species, by the similarity between parts of otherwise quite different animals, by the increasing complexity of the fossil record, and so on — Darwin stepped in to provide a key.

As we shall see, the scientific debate as to the validity of Darwin's new theory continued well on into the twentieth century. Indeed, there were moments when it looked as though it would be eclipsed by alternative viewpoints. But after *Origin* nothing was ever the same. Later debates argued only over the *mechanism* of evolution. Was Darwin right to pin his beliefs on natural selection, or was another process at work in nature? In such debates the principle of organic evolution itself was widely accepted. The question was no longer whether animal forms evolved, but how they evolved. For many people Darwin had turned evolution, hitherto only a speculation, into a fact, although they reserved judgment on whether the process was exactly as Darwin described.

We will consider these points again, but first we must return to history. We have seen how the scientists reacted to Darwin and we must now consider his reception among ordinary people, and within the church. For them the finer details of the mechanism of change were unimportant compared to the greater question of the nature of mankind.

6
THE CRISIS
OF FAITH

Onward, Christian soldiers!
Marching as to war,
With the Cross of Jesus
Going on before.

It was during the 1860s that Sunday school children first learnt to sing this rousing hymn of Christian warfare. Battle imagery has always been part of the language of the church. The apostle Paul used it to convey the Christian's struggle against evil. But in the 1860s battle cries were sounded with fresh vigour. In Britain the Crimean War and the Indian Mutiny, and in America the Civil War, lent a ready vocabulary of armoury and warfare. Many slogans derived from the American revivalist missions and Dwight L. Moody, who had preached to American soldiers, came to England in 1867 to lead evangelistic 'campaigns'. The struggle was for people's spiritual allegiance, but the language used echoed military idioms. From 1865 William Booth worked in the English city slums, calling his organization The Salvation Army. He based his regulations on real army codes; prayer was commonly referred to as 'knee drill' and the order 'fire a volley' meant shouting out 'Hallelujah'!

It is not surprising, then, that the arguments over Darwin's *Origin of Species* were couched in military language. The invective of Thomas Huxley, the sermons of ardent churchmen, the pamphlets of the popular press, all created a picture of science and faith at war over a doctrine of animal origins.

But this battle picture is a false one. There was certainly conflict, but the notion of science arrayed against the church is far too simplistic. Admittedly a certain whiff of gunpowder hangs over today's arguments between creationists and liberal evolutionists, but we must be careful of reading back into the past our understanding of the present.

The Bishop and the Bulldog

Darwin was lucky with the review of *Origin* printed in *The Times* newspaper. The selected reviewer admitted his incompetence to a friend, and the friend suggested Professor Huxley might help. Huxley later recorded, 'I wrote the article faster, I think, than I ever wrote anything in my life.' He was fulsome in his praise, and the 5,000-word review filled three and a half columns. He was not, however, uncritical, and though welcoming the new idea as an extremely valuable hypothesis he suggested that only time would prove whether it could carry the full weight of scientific evidence.

Huxley maintained this position throughout his life. His passionate support for Darwin was not for the minutiae of Darwin's theory, but for the tenor of his approach. He supported *Origin*'s naturalism, in contrast to any wanton invocation of miracles required by creationists. Darwin said at the end of *Origin* that the Creator might be behind his newly discovered laws, but Huxley was uncertain whether this was so. And since the question could never be answered with any certainty he preferred to ignore it altogether. He longed to be rid of explanations that talked only of 'final causes', that is, of references to the work of the Almighty. He wanted to find natural

The Salvation Army, set up in the nineteenth century, made full use of the military language with which Christians seemed so comfortable at that time.

causes, and these alone. Accordingly he ended his review:

> 'Mr Darwin abhors mere speculation as nature abhors a vacuum. He is greedy of cases and precedents as any constitutional lawyer, and all the principles he lays down are capable of being brought to the test of observation and experiment. The path he bids us follow professes to be, not a mere airy track, fabricated of ideal cobwebs, but a solid and broad bridge of facts. If it be so, it will carry us safely over many a chasm in our knowledge, and lead us to a region free from the snares of those fascinating but barren virgins, the Final Causes against whom a high authority has so justly warned us. "My sons, dig in the vineyard," were the last words of the old man in the fable: and,
> though the sons found no treasure, they made their fortunes by the grapes.'

Huxley saw the task of science as digging in the soil of knowledge. He expected progress to come through understanding and applying natural laws. The search for divine treasure was idle speculation.

Not many months were to pass before Huxley was publicly defending the new vine in the scientific garden. A significant test for the new evolutionary theory came in the summer of 1860. The British Association for the Advancement of Science met in Oxford for its annual meeting, and Darwin's new ideas were on the agenda. The debate between Huxley and Bishop Wilberforce on the last day of the meetings has now passed into scientific folklore as the greatest battle of the nineteenth century.

What exactly was said on that afternoon is uncertain. We have only personal reminiscences and the rather bland report in the *Athenaeum* magazine. A scholar from New York, Dr Draper, read a paper before the Association on 'The Intellectual Development of Europe Considered with Reference to the Views of Mr Darwin'. Draper was an important scientist: he claimed to have made the first photographic portrait by Daguerre's process in 1838, and took the first photograph of the moon two years later. His book, published in 1874, on the relationship between religion and science did much to spread the idea that the two disciplines were inextricably opposed. But that afternoon he simply bored his audience. Those assembled knew that the Bishop of Oxford would attempt to contradict Darwin's views, and they were anxiously awaiting the fireworks. Samuel Wilberforce had taken the trouble to be briefed on Darwin's ideas by the biologist Richard Owen. He was an able speaker, and had earned himself the nickname 'Soapy Sam'.

Huxley originally had no intention of attending. He knew of the Bishop's eloquence and saw no good reason to be 'episcopally pounded' on the Bishop's home ground. But Robert Chambers, now known as the author of *Vestiges*, by chance met Huxley in the street and persuaded him.

In the packed library of the Museum Dr Draper droned on for an hour. Eventually Henslow, in the chair, opened up the subject for discussion and the Bishop rose to reply. His speech was fluent and followed the lines of his comments on *Origin* in the *Quarterly Review*. Despite his cramming he was not sufficiently master of his facts, yet it was not in his biology but in his etiquette that he made the fatal blunder. According to one account he finished his speech by enquiring of Huxley whether it was 'through his grandfather or his grandmother that he claimed his descent from a monkey?' Huxley murmured to Sir Benjamin Brodie sitting next to him, 'The Lord hath delivered him into mine hands,'

A crucial debate about Darwin's theory took place in Oxford, where Thomas Huxley and Bishop Wilberforce of Oxford found themselves totally disagreed.

but refused to speak until the crowd was calling for him. Once on his feet he defended Darwin's views, but took advantage of the Bishop's distasteful jibe by adding that if the choice was between an ape for a grandfather or a man who misused his eloquence to introduce ridicule into a grave scientific discussion then he unhesitatingly affirmed his 'preference for the ape'.

The exact words, either of Wilberforce or Huxley, are now uncertain. Their effect is not. One lady fainted. The undergraduates cheered. Most of the audience applauded. To reply in such vein to a Bishop, especially in his own diocese, was rare indeed. The Bishop himself sensed that Huxley had won the day and did not rise again. The discussion continued, including

a contribution by Robert FitzRoy, now an admiral, recalling his arguments with the young Charles aboard the *Beagle*.

Perhaps in our day such an encounter might make the national headlines. At the time it caused a stir in Oxford and heartened Darwin's supporters. Outside of these circles it received little attention until the affair was reported in Darwin's *Life and Letters*, compiled in 1887 by his son Francis. Because the antagonists were an important bishop and a famous scientist, it has become a legendary symbol for the nineteenth-century battle of religion against science. Accordingly it takes its place alongside the trials of Galileo as an example of the supposed intransigence of the church in the face of progress. But, as with the Galileo affair, it is far from representing the whole story, and the debate that June in Oxford cannot be

The debate about evolution was inevitably personalized, as these pictures of Wilberforce and Huxley show.

described as the focal point of nineteenth-century intellectual struggle, as it is often made out to be.

Bishop Wilberforce had a first-class mind, and some of the criticisms voiced at Oxford he had already made in a review article on *Origin*. They were important objections which a number of eminent scientists were raising against the new theory. But of course it was not all science. He feared the moral and religious implications of natural selection, and he entered the debate with this in mind. But then neither was Huxley, or so it could be argued, as concerned with the minutiae of Darwin's theory as with its overall thrust. He welcomed *Origin* because it reflected his own philosophical conviction that everything happened according to natural law.

As the church waned in importance in Victorian England, and scientists established their own cultural importance, this Oxford battle became useful propaganda. Scienee had established its right to lecture the church. There were certainly deep tensions between Darwinism and Victorian church beliefs. But it is only because we are inheritors of Huxley's scientific culture that we remember this trouncing of a bishop as a major coup rather than a minor skirmish. The scientist David Lack has written:

'There is no need to repeat here either the ill-founded attacks of the churchmen on Darwinism, which have often been recalled, or the ill-founded attacks of Darwinists on the Church, which are usually forgotten. It is enough to say that, while many churchmen, both Roman Catholic and Protestant, showed a lamentable ignorance of the findings and the principles of biology, the same could be said of various Darwinists in relation to theology. Mixed up with the truth, there were ignorant, unjustifiable, absurd and violent assertions on both sides, and it is perhaps through the spirit of the age that we remember the arrogance of the conservative theologians rather than of revolutionary Darwinists.'

LOUIS AGASSIZ

The main spokesman in America against Darwinism was the Swiss-born naturalist Louis Agassiz. During the spring of 1860 he publicly debated with Asa Gray the validity of Darwin's arguments. At the time Agassiz was the better-known, although now the greater scientific reputation rests with Gray.

Agassiz was thirty-nine when he moved to America. He had studied under the famous Cuvier, and had already published his revolutionary theory on the Ice Age. American science had none of the prestige of the European academies, and Agassiz moved to the new continent not out of a pioneering spirit but because he was penniless. Printing his work on fossil fishes had exhausted his small salary as Professor of Natural History in the Swiss-Prussian town of Neuchatel. The attraction of a lucrative lecture tour made him embark for America, and he picked up what English he could from conversations with the ship's captain.

Americans were excited about his coming, and turned out in their thousands. In less than six months he had earned nearly $6000. He stayed. And he imbibed the pioneering spirit. In 1850 he wrote, 'The time has come when American scientific men should aim at establishing their respective standing without reference to Europeans.' Seven years later he proved his words were genuine when he refused an important Chair in Palaeontology in Paris.

Agassiz did much to establish American science. He had valuable contacts with Europeans, he used his own prestige to start projects and influence businessmen, and he appeared regularly on the lecture platform enlisting his audiences in the support of science. On his death in 1873, hundreds of 'Agassiz Associations' sprang up across the country to encourage young people to study natural history.

But he never understood Darwinism. Too entrenched as a disciple of Cuvier, he continued to maintain that species were divinely created and fixed for all time. Darwin's theory was only 'conjecture'. The progression of fossil forms from one strata to the next reflected the developing plan of divine creation, not natural transitions under pressure from the environment.

Essays and Reviews

Huxley might enthuse over the value of natural knowledge, but Victorian England was not yet ready to abandon the revelations of the Bible in favour of the researches of science. There was an inevitable clash with Darwinism over the history of origins as (apparently) given in the Bible. We have seen how this became an important issue when geologists first proposed an ancient date for the formation of the world. The debate over *Origin* added nothing new, but it did coincide with a major upheaval in theology which gave the impression that the new science was the cause of the infidelity. During the later part of 1860, Victorian England rocked with outrage at the views of seven essayists in the publication *Essays and Reviews*.

The contributors were all clerics, save one. Their intention, as one of them expressed it, was 'to say what we think freely within the limits of the Church of England'. The subjects covered were broad and their quality patchy, but it was the spirit of the book which caused offence. The contributors doubted whether the Bible could be taken, letter by letter, as the word of God. They believed that it needed to be understood in its original historical and literary setting, much as any other book. Such ideas had been commonplace in German universities for a number of decades, but to an essentially conservative church in Britain they came as a deep shock.

Almost four decades later, the Church of England was to enthrone Frederick Temple, one of the essayists, as archbishop. He firmly believed the Creator had endowed his original creation with all the necessary powers to evolve into a myriad of living creatures. The Rev. Brownjohn tried to stop Temple's enthronement, for he had himself resigned his parish as a result of his own conversion to evolution. But by then a bishop who took evolution as axiomatic was no longer a scandal. Simple worshippers still hung tenaciously to a literal Bible and a firm separation of human beings from the apes, but although belief in evolution was not yet universal

it was, in most circles, quite acceptable.

Back in the 1860s, however, it was quite another matter. This readiness to treat the Bible in a literary, rather than a literal, way was not acceptable from the outset. Instead there was fierce opposition. Bishop Wilberforce said the writers could not 'with moral honesty maintain their posts as clergymen of the established church'. Two of the essayists were brought before the ecclesiastical courts, and a national declaration on the inspiration and divine authority of the Bible was signed by over 10,000 clergymen.

In England at least, the arguments over evolution became inextricably linked with the rise of biblical criticism. In the popular mind Darwin's theories were another blow to orthodoxy and full belief in the Bible.

At heart the two controversies were similar. Darwin claimed that animal structures, once devoutly ascribed to the miraculous power of God, were the outworking of regular scientific laws. The authors of *Essays and Reviews* believe the Scriptures were the product of ordinary human writers. Darwin's naturalism in nature was matched by the cleric's naturalism in Scripture.

At a deeper level, of course, there was no religious infidelity. Darwin was pleased when people saw his laws as the instruments of God, and the essayists still believed God spoke through the Bible. God's Spirit infused the one as he did the other. But both *Origin* and *Essays and Reviews* wanted to limit the miraculous. Animals were not 'directly fashioned by God' any more than the Bible was directly dictated.

In the months immediately after the publication of *Origin* more newspapers and magazines came out against the theory than in support. The low-brow or popular dailies were firmly set against the new ideas, while the thoughtful magazines were more accommodating. Acceptance spread slowly. But historical research has shown that it was the idea of evolution, and not the mechanism of natural selection, which gained credence. Whereas twenty-five years earlier evolution could not be accepted because there was no known mechanism, now evolution was acceptable even though Darwin's own suggested mechanism was believed to be inadequate. Somehow people were more ready to believe the overall idea once they had been given a plausible theory. And when the theory crumbled they still clutched at the idea!

The historian Owen Chadwick has written of this period:

'The public accepted the doctrine of evolution for a bigger reason than the simple probability established by Darwin, namely that, if they did not, their mental picture of the origins suddenly became an intolerable blank.'

With institutionalized religion growing weaker, people still needed an explanation of their origins. And in the empire-forming days of Queen Victoria they liked a creed which spoke of a future prosperity towards which they were progressing, rather than the Bible's picture of a past Garden of Eden from which they had been banished.

There was, however, a continuing objection to mankind's inclusion in the scheme. People became less perturbed by the thought that animals evolved, but they did not like to think of themselves as distant cousins of primeval protozoa.

Descended from apes

On hearing of Darwin's new theories, the wife of the Bishop of Worcester is reputed to have exclaimed:

'Descended from apes! My dear, let us hope it is not so; but if it is, that it does not become generally known.'

But generally known it became. It was the linking of humanity with the apes and the subsuming of all under natural law that gave most offence. In 1861 the controversy was fuelled by exhibits in London of stuffed gorillas. Widespread knowledge of the existence of gorillas went back to 1847, but apart from

At the time when 'Origin of Species' was published, gorillas had only recently become widely known. The points of resemblance to human beings were striking.

explorers no European had ever seen one and Paul Belloni du Chaillu's exhibition caused a sensation.

Du Chaillu was an American of French extraction who had lived as a child in West Africa. A grant from the Philadelphia Academy of Natural Sciences had enabled him to return and explore. He claimed the gorillas were the results of his own shooting, but was later accused of being a charlatan since forensic evidence did not match his much-vaunted claims of encounters with the beasts. But whatever the genuineness of du Chaillu, the gorillas were real enough. In London impostors were nothing new but the 'missing link' certainly was. So the crowds flocked to see. The *Daily Telegraph* voiced its concern, fearing that the answer as to whether or not the gorillas were allied to the human species might 'have disastrous consequences to the national peace'.

The problem, it seems, was that du Chaillu

These illustrations of ape skeletons appeared in Thomas Huxley's important book, 'Man's Place in Nature'.

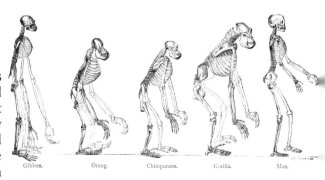

Gibbon.   Orang.   Chimpanzee.   Gorilla.   Man.

NEANDERTHAL MAN

In 1856, the bones of a strange man were found in a small cave near Düsseldorf. The cave, part of a quarry, was in the steep-sided Neander valley, and workmen digging for limestone reported that they had unearthed some old bones. The quarry owner took the remains to a local scientist, Johann Fuhlrott, who in turn sought the advice of an expert on anatomy, H. Schaaffhausen. The stocky bone structure and prominent ridges over the eyes was evidence, Schaaffhausen declared, of the 'most ancient races of man'.

Others were not so sure. Even if the growing ideas concerning progressive organic development were true, most scientists were convinced that humanity itself was of very recent origin. And of course this was a single find. Another 'Neanderthal' skull had been found in Gibraltar eight years earlier but it was not recognized as such until debate on the German bones aroused awareness that such skulls existed. The Gibraltar skull simply languished in a small museum of natural curiosities. When it was rediscovered and exhibited at the British Association for the Advancement of Science meeting in 1864 it helped to confirm our human links with the past. The skull had ape-like characteristics but, as Huxley warned in his 1863 book *Man's Place in Nature*, 'In no sense can the Neanderthal bones be regarded as the remains of a human being intermediate between men and apes'.

Many equally eminent scientists examined the Neanderthal bones and pronounced that they belonged to a diseased human rather than to an ape-like ancestor. The poor creature had suffered from rickets as a child, said the Professor of Anatomy at Bonn University. Alternatively the curved leg bones suggested life in the saddle, so the professor also pondered whether the supposed ape-man had in fact been a deserter from a Cossack army which, earlier in the century, had camped nearby. Of course the presentation of the Gibraltar skull removed this last rather fanciful explanation (unless, that is, Cossacks had been deserting all over Europe) but for many years to come idiocy and rickets was a common alternative explanation. The next fossil man was not to be discovered until the 1890s.

In 1887 Eugène Dubois resigned his lectureship at the University of Amsterdam and set sail for Sumatra. He was determined to find the 'missing link'. Wallace, who had come to the idea of evolution by natural selection at much the same time as Darwin, believed the Malay archipelago held the secret to mankind's ancient past. And Dubois believed Wallace. The rest of the scientific community thought the whole scheme madness, for surely Europe with its

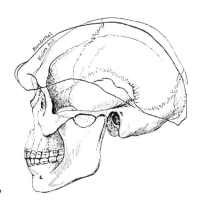

A particular stage in human evolution is known as 'Neanderthal' after an ancient skull found in the valley of the Neander River in Germany. The skull in this picture was found in Western Port, Australia. The contours of the Neanderthal skull are superimposed.

finds of Neanderthal man was the source of humanity?

Dubois proved them wrong, or at least so he thought. In 1891 he discovered a fossil skullcap which he first described as belonging to a chimpanzee. But then, the following year, he found nearby a thigh bone, clearly human. Linking this second find to the first, Dubois revised the classification of the skull. He had found, he believed, an ape-man, and called it *Pithecanthropus erectus* ('upright ape-man'). Ernst Haeckel, in drawing up his evolutionary trees, had first suggested *Pithecanthropus* as the name for the animal part-way between man and ape. Now 'Pithecanthropus' had walked out of the Sumatran jungles.

Dubois telegraphed home that he had discovered the missing link and brought his specimens home to show in one scientific congress after another. His fellow-scientists warmly applauded his efforts, yet strongly doubted his conclusions. Some questioned whether the skull and the femur belonged to the same individual, others believed the skull was of an ape, others of a human. Hardly anyone believed he had found an intermediate ape-man. Dubois vigorously defended his cause, but in the end, out of pique, he withdrew his specimens from the sceptical gaze of science and locked them in the safes of the Teyler Museum at Haarlem, his home town. There they remained for thirty years until comparisons with other skulls found in China showed that *Pithecanthropus* was an early, but true, hominid. It was far above the stage of an ape and could not be considered as an example of an animal midway between mankind and the lower animals. It was later renamed *Homo erectus* by Ernst Mayr. Dubois died in 1940 a disappointed man. The missing link was still missing.

spoke chillingly of the ferocity of the gorilla in the wild. This was pure fantasy, but his audience was not to know that. So, taking du Chaillu at his word, they sensed that mankind's more immediate ancestors were not exactly the Victorian gentlefolk they would have wished for. And what did that say for present-day morality? Was humanity any more than a tamed beast?

The newspaper article continued:

> 'Even if Mr Darwin and his friends could persuade us that our distant ancestors were guinea pigs or caterpillars, people would not, we are inclined to think, have founded a new system of ethics on the discovery. We should still be more interested in our present and our future than in our past. But human dignity and human feeling both revolt against the absurdities of the would-be scientific men. God made man in his own image, says the Book of Books; and though this is ancient testimony to the divinity of our origin, it has not yet been upset by modern conjecture or modern assumption.'

But the *Telegraph* was wrong. People were both affronted as to their dignity and challenged with respect to their ethics. If man was only a brute beast then only as a brute beast could he be expected to behave. If natural selection implied savagery in the wild, did not the same biology justify immorality within human society?

Sinking into degradation

This question of the brutalization of humanity lay behind much of the fear in Germany of introducing the new science into the classroom. *Origin* was first translated into German by the palaeontologist Heinrich Bronn. It was not a happy translation, being far too literal and laborious. Bronn also added an appendix of objections which only indicated that he had misunderstood much of the argument. Darwin was therefore pleased when in 1866 an admirer, Viktor Carus, offered to undertake a new translation. The revolutionary theory made quick inroads into German science, although the converts tended to be young men at the start of their profession rather than the established elite. Darwin wrote to a correspondent, 'The support which I receive from Germany is my chief ground for hoping that our views will ultimately prevail.'

The public debut of Darwinism in Germany came with a speech by the youthful zoologist Ernst Haeckel at the annual conference of the Association of German Scientists and Physicians in 1863. Haeckel was filled with the enthusiasm of a new convert and, like most converts, made exaggerated claims for his new faith. Darwinism, he said, afforded an entirely new perspective on human knowledge. It spoke of change and progress. Here was a message of liberation, with lessons to be learnt from the struggle in nature.

> 'Progress is a natural law that no human power, neither the weapons of tyrants nor the curses of priests, can ever succeed in suppressing ... Standing still is in itself regression, and regression carries with it death. The future belongs only to progress!'

From the outset in Germany, Darwinism was adopted as an ideological tool which presaged a new future. As new animal species had risen from the graveyards of the old, so modern social reforms would eventually triumph over political conservatism. The thwarted revolutions of 1848 had created a pressure for liberal reform. Where overt fighting was impossible, propagation of materialistic science was not. The reformers hoped to loose the chains of authority and superstition by mass education. In the 1850s, hopes were focused on philosophical materialism.

The recently discovered principle of the conservation of energy implied that although energy or matter changed, it was never created or destroyed. More than the ideas of natural selection, it was this doctrine which was destructive of an interpretation of nature which saw God intervening through miracles. If energy

The 1848 revolution in Germany showed a desire for social progress. Ideas of social evolution were a potent weapon in the hands of materialistic philosophers.

only changed in form (and not in amount) then there was no sign of God's infusion of power through miracles. Ludwig Büchner published *Force and Matter*, which stressed the eternity of matter and the absence of any spirit or God. Darwin's theories, with their apparent independence of divine design, were welcomed as allies.

Haeckel both helped and hindered. He did a very effective job in popularizing Darwin's thought (while *Origin* remained largely unread). But by using it as a weapon against religion, and in particular Christianity, he provoked counter-attacks in the name of faith and traditional morality. The issue surfaced in the teaching of Hermann Müller, a biology teacher in the *Realschule* in Lippstadt. In 1876 the *Westfälischer Merkur* accused Müller of subverting religion by teaching evolution. And religion, they pointed out, was one of the main pillars of the social order. Someone sent the article to the Prussian cultural minister and although he appeared satisfied with Müller's defence it brought the issue into the open. Some conservative papers took no trouble to confirm the facts, denouncing Müller as a corrupter of the youth and demanding his removal. Müller in return sued for libel, and the affair wound its way through the courts.

There was a real fear that a consequence of such teaching would be to raise 'a generation whose confessions are atheism and nihilism and whose political philosophy is communism'.

Biology was feared as a springboard for political action. The progressives saw the new theory as symbolizing their own struggle to be free from oppressive authority — whether from the state or from the church. At stake was not the veracity of a theory in biology, but the stability of society. And such stakes do not allow an impartial weighing of the evidence!

This was the real crux of Darwinism. It demanded radical rethinking of fundamental attitudes. In Britain, Darwin's old professor of geology wrote to him with much pain in his heart:

'There is a moral or metaphysical part of nature as well as the physical. A man who denies this is deep in the mire of folly. 'Tis the crown and glory of organic science that it *does* through *final cause*, link material and moral . . . '

Professor Sedgwick believed that once science ceased to attribute the laws of nature to the will of God, but rather set them up as independent causes, then all sense of a divine plan was lost. And without this overall plan, humanity too was set loose to find its own way in the world. So-called 'natural' laws should be understood as linked to divine power, much as today we would explain a car journey in terms both of car mechanics and the direction chosen by the driver. If there is no one at the wheel, then an accident is sure to follow. The professor continued,

'You have ignored this link; and, if I do not mistake your meaning, you have done your best in one or two pregnant cases to break it. Were it possible (which, thank God, it is not) to break it, humanity, in my mind, would suffer a damage that might brutalize it, and sink the human race into a lower grade of degradation than any into which it has fallen since written records tell us of its history.'

There was conflict over evolution in the nineteenth century, but it was never a simple battle between religion and science, between Bible literalists and evolution progressives. Disagreements over the Bible formed only a small part. The danger in Darwinism did not lie in its desire to investigate the natural world, for such investigation was fundamental to the understanding of science forged by the Puritans in the seventeenth century; they delighted in discovering God's ways and in reading the book of creation. No, the danger lay in the fact that the nineteenth-century mind wanted

HAECKEL AND BÖLSCHE, PRIESTS OF DARWINISM

As used to be said of Kings Saul and David, if with his books Haeckel slew his thousands then his friend Wilhelm Bölsche slew tens of thousands. Their conquests were over the 'philistines' who still believed there was a spiritual force behind the universe. And between them their popularization of a materialistic philosophy bolstered by Darwinism convinced generations of Germans that God was no longer to be believed in.

Haeckel began by offering a popular account of Darwin's new theory. His *Die Natürliche Schöpfungsgeschicte* of 1868 (later translated into English by Sir Ray Lankester as *The History of Creation*) traced the natural descent of life. But in areas where Darwin held back, on the origins of humanity and on the initial creation, Haeckel was prepared to extend the web of natural cause and effect. And he also saw evolution as more than a biological theory. It was a whole way of picturing the world, driving out the necessity for belief in a guiding Spirit. Evolution spelt progress, and the demise of God spelt the end of the authority of reactionary church leaders and institutions. Haeckel always claimed he was not a materialist. Although he rejected any view which believed mankind had a distinct spirit or soul in addition to a physical constitution, he saw life and consciousness as built into the very particles of the universe. Matter, even the simplest, was not inert. In its rudimentary way it possessed 'soul'. In his most famous book, *Riddle of the Universe*, he claimed that plants were conscious.

To all this Bölsche added poetry. A poetry of wonder in the face of the mysteries of creation. Creation was turned into romance. He opened his 1898 book *Love-life in Nature: The Story of the Evolution of Love* by transporting his readers to 'a lovely spot — East of San Remo'.

'Strata of stone, that once were soft ocean bottom millions of years ago, crop out of the soft green contour of the coast like a phantastic citadel. The blue Mediterranean exposed them, wearing them away, not with a rough fist, but through an infinite length of time touching them over and over again as in a dream with delicate white foamy hands . . . At the outermost rim of the cliff the foamy coronet of the free onrushing waves flashes incessantly like dazzling white wings that close and spread fanlike in the sun. Beyond, far as the eye can see, all is blue; deep and bewitching blue . . . '

The title of his book was no accident. Bölsche played down the English insistence that nature was violent and progressed through struggle. Instead he saw co-operation and symbiosis. Where he had to acknowledge the brutal side of nature he preferred to think this was a transitional stage paving the way to a greater harmony. It was total harmony which nature was struggling to achieve, and

increasingly to interpret its findings without recourse to God.

Huxley, as Darwin's bulldog, was not anti-religious. But, as we have seen, he wanted to limit scientific explanation to observable cause and effect. He wanted no truck with Final Causes. Yet a science without final causes implied a world without God. And if there was no creator-God then human beings were indeed no more than the beasts. It was a false equation, but it was no easy matter to hold biology in a separate compartment from morality. Indeed, the natural theologians from Paley onwards had taught them not to. Then, as now, an understanding of how we arrived in the world seemed to impinge on beliefs about how we should act in it.

Reconciliation

During the critical years of the Darwinian controversy Asa Gray (1810-88) was professor of natural history at Harvard in the United States. He was a first-rate scientist, writing numerous botanical works, including the massive *Synoptical Flora of North America* (1878). Darwin turned to him for botanical information, and entrusted him with details of his theory in the years before publication. When *Origin* was published in England, Gray did his best to limit the pirate copies printed on his side of the Atlantic and produced an authorized edition through D. Appleton and Company. He was also a devout and orthodox Christian, of evangelical persuasion, and his arguments reconciling natural selection and Christian belief so pleased Darwin that he had

---

he echoed Herbert Spencer in believing that the cosmos was moving from chaos to cohesion. As amoebas had evolved into mankind, so primitive societies had moved through Greek civilization and on up towards the heights of German culture.

Bölsche knew he was providing a new world-view.

'And from the depths of the human soul, whence also the lessons of the gospels have come, still another voice whispers into my inner ear, a voice first heard in the wisdom of the ancient Indians. And it says that the band of community and brotherhood is not limited to man, but that it encompasses all things on this earth, all things that grow up and evolve to their peak under the sun's rays and in the silent grip of holy universal laws.'

It was a view eagerly adopted by the German middle- and working-classes. Here was hope. True, it was ill-defined and focused on no clear political programme. But it captured people's imagination in a way that the more complex Marxist ideology could never do. It undermined church and

state, those apparent denizens of oppression. It offered comfort that, one day, all would be well.

Popular science proved a powerful weapon. Haeckel, Bölsche and others all wrote in opposition to the church. Alfred Kelly, who has studied the rise of popular Darwinism, has written:

'Without a Christian genre of popular science, Christians missed the opportunity to give the Darwinians some of their own medicine.'

If Bölsche's writings were used by the political left then Haeckel's could be taken as supporting the right. His answer to everything was more and better education which would allow people to throw off the shackles of a reactionary state and church. If anything, the Monist League established in 1906 to promote Haeckel's philosophy was politically left of centre. Bölsche was one of its members. But Haeckel talked of the superiority of certain groups and races and the inevitability of struggle:

'The theory of selection teaches that in human life, as in all animal

and plant life everywhere, and at all times, only a small and chosen minority can exist and flourish, while the enormous majority starve and perish miserably and more or less prematurely.' *Freedom in Science and Teaching*

Alongside popular Darwinism there was a growing, but separate, movement supporting German nationalism. This movement could draw on evolution and ideas of the supremacy of the few, but it did not arise from Darwinism itself. At times the two strands coalesced, as in the writings of Alfred Ploetz who preached that to safeguard the future of the race it was necessary to kill unsuitable children at birth, much as nature weeded out the unfit. But the route from Haeckel to Hitler is not a single line of descent. Hitler employed a crude biology which emphasized survival of the fittest and the need for racial purity, but he plundered Darwinism for his own racial purposes. His Nazi Propaganda Ministry failed to list Haeckel's books as recommended reading.

them reprinted as a pamphlet which he sent to his scientific colleagues.

How could Gray accomplish the reconciliation between science and theology which others believed impossible?

There were two parts to the reconciliation:

☐ *First, Gray understood so-called 'natural' laws as God's instruments of creation.* What we term laws are in reality our observations of God's regular, and providential, activity. Gray, therefore, had no quarrel with natural selection. Darwin had noted the mechanism which engineered evolution; Gray simply added his belief in the Engineer behind observable nature. Darwin had done for biology what Newton had done for mechanics. They had both sought out the laws governing movement, and the eye of faith had as much right to claim that this was a discovery of God's ways of working as the atheist to deny it. Some Christians had bemoaned the fact that Victorian science had slipped into a search for miraculous interventions. They believed the task of science was to discover the regular working

Professor Sedgwick, Darwin's old teacher of geology, believed his pupil was wrong to neglect God's control over nature.

of God in nature and they accordingly received Darwin's ideas with enthusiasm. Rather than seeing *Origin* as harmful to religion, they believed it restored a truly Christian perspective in which science uncovered God's activity in the natural world. Charles Kingsley, English cleric and noted author, wrote to Darwin:

'I have gradually learnt to see that it is just as noble a conception of Deity to believe that He created primal forms capable of self-development into all forms needful . . . as to believe that He required a fresh act of intervention to supply the *lacunas* which He Himself had made. I question whether the former be not the loftier thought.'

☐ *But second came the issue of chance.* This was a more difficult area of Darwinism for Asa Gray to reconcile with his Christianity. Natural theologians from Paley onwards had taught the Victorians to see organic structures as examples of God's design. Just as the intricate design of a watch leads you to postulate a watchmaker, so the complexity of, say, the human eye should lead one to belief in God. The earlier debates in geology had challenged the harmony between science and a literal Bible, and had questioned

This Punch cartoon of Darwin and an ape with a mirror has become the best-known humorous presentation of the popular image of Darwinism.

God's intervention in miracle. The last bastion of proof of God's presence in the world was thus the design of natural organisms. Show that these were not designed, but only the products of time-plus-chance, and belief in God was thrown into question.

Darwin never understood why one animal varied from another of the same species. But vary they did, and nature, through natural selection, weeded out those variations which were not beneficial. But the variations were not designed with the future prospects of the animal in mind; they simply happened and were found to be either advantageous or harmful. A builder constructing a stone wall selects the stones which suit his purpose, but no one could claim that the shape of each stone was somehow determined by nature for the purpose of forming walls. The appearance of design (the finished wall) is therefore a product of a law (the builder or, in nature, natural selection) and the availability of plenty of chance variations. This, for Darwin, was the crucial point and the source of his disagreement with Gray. Both acknowl-

This cartoon points to an idea which many still hold, that particular views about human origins can somehow reduce the value of a human being.

MAN·IS·BVT·A·WORM·

edged ignorance as to the source of variations, but whereas Gray believed God's hand was at work in shaping them for a particular use, Darwin did not.

Because the origin of the variations poured into nature's sieve was unknown, Gray could hold that they had been 'led along certain beneficial lines'. God had, so to speak, loaded the dice so that certain combinations would eventually arise. In a picturesque analogy, he saw natural selection as the rudder of the ship of evolution, determining its movement, while God-inspired organic variations were the driving wind. If the wind direction changes every few minutes, sailing any distance is impossible. So Gray doubted whether natural selection operating on truly random variations would ever have produced the complexity and design in organisms we observe today. A mechanism of pure chance was insufficient. And so he saw God directing each and every variation for his own ends.

In the end both positions rest on faith. The non-believer sees the variations as completely random with respect to their final use, yet marvels at the power of a process such as natural selection to produce both beauty and complexity. The final animals have the appearance of being designed. But in reality they are only with us now (and we too are only alive) because these structures were able to survive while others died out. In contrast, the believer sees beneath the randomness the overarching purposes of God. Either God is in some way directing the source of variations (as Gray believed) or the whole process of chance-variations-plus-selection is a carefully thought-out agency for building creation.

The non-believer finds he has a cruel creed. What he counts as beautiful or significant is no more than a complex series of accidents welded together into a surviving whole. Millions upon millions of other possibilities might have been, but pressures of survival gave those 'designs' the thumbs down.

In contrast, the believer finds comfort in design and intelligence behind the universe.

QUESTIONS TO CHARLES KINGSLEY AND TO ASA GRAY

In their reconciliation of Darwinism with Christianity, both Charles Kingsley and Asa Gray used two concepts that could themselves be questioned. The first relegated God's creative activity to an initial moment of creation, and the second looked for God in what was inexplicable. Both approaches were repeated by others, but neither proved satisfactory.

In Kingsley we should note the use of self-development. He held that God originally gave to his creation an in-built capacity to develop along certain lines. Using an analogy not available to Kingsley we could say that animals, like computers, are 'programmed' to evolve. Once the programme statements have been inserted the programmer can sit back and watch his creation unfold. This is a deistic approach and was echoed by Darwin in the final lines of *Origin* when he talked of powers 'having been originally breathed by the Creator into a few forms or into one'. It is not orthodox theism, which prefers the picture of a car driver to that of a computer programmer. Whereas the programmer types in his command statements at the beginning and waits for the programme to run, the car driver is constantly altering the speed and direction of his vehicle. Each movement results from the continuing activity of his mind, not from pre-programmed requests. The more orthodox Christian viewpoint is to see natural law as a continuing statement of God's activity in the world. If our laws are constant from one day to the next that is not because they were 'determined' on the day of creation, but because God is faithful day by day now. He is no armchair creator. He

creates and sustains everything.

Nevertheless there were many Christians who understood evolution along fundamentally deistic lines, and saw God's creative role as limited to his initial act of creation. It is a view which does not do justice to the creation story in Genesis, since those chapters have more to do with acknowledging God's present sustaining power than with recording an initial creative act.

Deism was not Asa Gray's error. Gray may be accused of believing in a 'God of the gaps'. He argued that God was required to direct the course of evolution for completely random variations would never lead to advancement. Helpful variations would be cancelled by unhelpful ones, and no progress would be made. And since naturalists did not know the source of these variations, they must be the hidden means by which God directed evolution. God controlled the

Charles Kingsley, a Christian social reformer and novelist, tried, as Asa Gray did, to bring together Darwinism and Christian faith.

millions of minute variations for his own purposes. Gray's logic appears to be that since the source of variations is inscrutable, then it must be divine.

Too often people make the claim that where science cannot fathom,

there God is at work. Gray opened himself up to the danger that if later scientists discovered the mechanism of variations, and showed them to be random indeed, then his picture of the providential driving wind of God would look superfluous. And this is the very claim of Jacques Monod and Richard Dawkins, scientists of our generation. Using Paley's picture of a divine watchmaker, Dawkins has called one of his books *The Blind Watchmaker*. His belief is that there is no divine hand behind organic variations. They occur quite randomly, and natural selection is the powerful, but natural, process by which complexity is built up.

But this is not the end of a Christian interpretation of Darwin. The idea that variations occur quite randomly need not mean that 'everything is here by chance' and there can be no God. Many physical processes are made up of a series of random movements, which when added together produce an effect none the less predictable. The air molecules around your head are buzzing in random patterns, yet taken as a whole they follow the physical laws of pressure and compression. Could not organic variations, quite random with respect to the final evolved structure, be added together through natural selection in much the same way? The final design is derived by the law of natural selection acting on countless variations, and the Christian is quite prepared to see law as a statement of God's activity. In fact, the very randomness at the heart of the process may be a way of exploring different possibilities in evolution, much as a chess computer selects the best move by exploring the outcome of countless other moves. Further comment must await a later chapter, but it is always open to the believer to see the creative hand of God working through any scientific process.

But he or she is left with questions as to the reason why God used this apparently wasteful and clumsy process. If God directs each and every change why do variations occur that are injurious? For every advanced organism are there not countless more that die? So why has God designed a process which requires the strong to usurp the weak? Though the believer gazes in wonder at the ultimate divine design, he also asks whether the process involved shows God to be a loving benefactor or a clumsy mechanic.

In the debate among the geologists thirty years earlier the central issue had been the nature of God. How did he operate in the world — by regular law or by miraculous intervention? Those, like Lyell, who advocated gradual change, still saw science as revealing the handiwork of God. Indeed they wondered at the marvellous design in nature and praised the Great Designer. Then came Darwin. He explained such an intricacy as the human eye as the product of chance variations collected over time. This took the argument one crucial stage further. Even if the scientist still acknowledged God as creator, how could God be said to work through the savage and wasteful mechanism of natural selection?

Darwin's loss of faith

There is a fanciful story that right at the end of his life Darwin came back to a Christian faith. After expressing his love of the Bible

Asa Gray

A party of naturalists make camp in the Rockies in 1877. To the left, seated, is Sir Joseph Hooker. Professor Asa Gray is to his right.

he reputedly confessed to a visitor called Lady Hope:

> 'I was a young man with unformed ideas. I threw out queries, suggestions, wondering all the time over everything and to my astonishment the ideas took like wildfire. People made a religion of them.'

People undoubtedly made a religion of Darwin's ideas. But accrediting to Darwin himself an orthodox faith is pure fabrication. During the *Beagle* sailing he at first hoped to return to take up a career in the church. Indeed he was sometimes mocked by the crew for quoting

In his later years Darwin was troubled by an unresolved conflict in his mind between the beauty and grandeur of the world and the cruelty of some natural processes.

the Bible. But he records in his autobiography how he came to doubt the veracity of the Bible and so 'gradually came to disbelieve in Christianity as a divine revelation'. He drifted from orthodoxy to liberal beliefs in God, and from thence to an agnosticism which at times clarified into atheism.

> 'The old argument (for the existence of a personal God) from design in nature, as given by Paley, which formerly seemed to me so conclusive, fails, now that the law of natural selection has been discovered. We can no longer argue that, for instance, the beautiful hinge of a bivalve shell must have been made by an intelligent being, like the hinge of a door by man. There seems to be no more design in the variability of organic beings and in the action of natural selection, than in the course which the wind blows. Everything in nature is the result of fixed laws . . . '

Darwin's God became veiled in a web of natural laws. And when he considered the problem of suffering, the 'suffering of millions of the lower animals throughout almost endless time', he doubted the benevolence of the cause behind the universe. Yet he wanted to draw back from complete unbelief, for he sensed 'the extreme difficulty or rather impossibility of conceiving this immense and wonderful universe . . . as the result of blind chance or necessity'. He therefore still called himself a theist.

Darwin was not alone in his struggle. To accept the theory of natural selection, as propounded in *Origin*, was to sail between two dangerous rocks, either of which could shipwreck faith. On the left stood the granite peak of purposelessness in nature, prepared to sink the Victorian belief in divine design. On the right towered the fearful question of evil and a God of love. Could God really have designed a cat to play with a mouse, or planned for the death of countless organisms to be the ladder by which mankind ascended to the divine image?

The poet Tennyson had passed through this narrow channel a decade before Darwin published. His poem *In Memoriam*, written in memory of a lost friend, first bequeathed to us the phrase 'nature, red in tooth and claw':

> Are God and Nature then at strife,
> That Nature lends such evil dreams?
> So careful of the type she seems,
> So careless of the single life;
>
> That I, considering everywhere
> Her secret meaning in her deeds.
> And finding that of fifty seeds
> She often brings but one to bear,
>
> I falter where I firmly trod,
> And falling with my weight of cares
> Upon the great world's altar-stairs
> That slope thro' darkness up to God,
>
> I stretch lame hands of faith, and grope
> And gather dust and chaff, and call,
> To what I feel is Lord of all,
> And faintly trust the larger hope . . .
>
> And he, shall he,
>
> Man, her last work, who seem'd so fair,
> Such splendid purpose in his eyes,
> Who roll'd the psalm to wintry skies,
> Who built him fanes of fruitless prayer,
>
> Who trusted God was love indeed
> And love Creation's final law
> Tho' Nature, red in tooth and claw
> With ravine, shriek'd against his creed —
>
> Who loved, who suffer'd countless ills,
> Who battled for the True, the Just,
> Be blown about the desert dust,
> Or seal'd within the iron hills?

The Victorians always thought of Darwin's theory as underlining this savage view of nature. In some respects it did, but he also talked of 'sexual selection' where the ability of an animal to reproduce was more important than its skill at hunting. Some organisms obtain their advantage by producing more offspring rather than by consuming their enemies.

Tennyson's poem 'In Memoriam' wrestles with the apparent contradiction between faith and the pain and suffering in nature.

But even with this amendment nature still seemed cruel and wasteful. Asa Gray maintained his Christian faith by stepping back to consider the whole canvas of evolution. A close view of an impressionist painting shows only a bewildering collection of dabs of paint, but as you step back the beauty of light and shade, form and design appear. So, Gray argued, the individual ways of nature may seem strange

'Nature red in tooth and claw' is seen vividly among the predators. Darwin found it hard to reconcile belief in God with the cruelty that natural selection implied.

but the overall design wonderfully leads up to mankind. If there is cruelty and waste it is a necessary sacrifice for the greater glory of the higher animal forms.

Not everyone could accept Gray's solution. And it was the search for both purpose and goodness in creation, that made some scientists and theologians abandon Darwin's biology. They turned to construct a creed more to their liking.

Spencer and the American Dream

As the nineteenth century wore on Darwin's theory of natural selection continued to be criticized. Yet belief in evolution became established, despite the lack, as already mentioned, of a mechanism by which natural selection might work.

To the most casual observer Victorian industrialism was changing the face of the land and (for the lucky) the wealth of the workers. Progress was evident, and a creed was necessary to express it. Darwin's natural selection reflected the struggle inherent in nineteenth-century capitalism, but it did not capture its glorious sense of progress. And so people turned to Herbert Spencer.

Spencer was born in 1820, the son of a Derby schoolmaster. A man of great intellect, he poured out philosophical works almost to the end of the century. At the core of each work was the single explanatory idea of evolution. In some ways he may be said to pre-date Darwin, for he had established the broad outlines of his doctrine well before 1859. He came close to discovering natural selection, and Darwin used Spencer's phrase 'survival of the fittest' in later editions of *Origin*. But whereas Darwin's work was based on observations of nature, collected together into a scientific hypothesis, Spencer's was an all-embracing philosophical idea buttressed by any examples he could turn to his advantage. The zoologist Ernst Mayr has said,

'It would be quite justifiable to ignore Spencer totally in a history of biological ideas because his positive contributions were nil.'

But his ideas suited the Victorian mind. And they were nectar to the still-developing American society.

Spencer claimed that the movement of all things was towards increasing complexity and interdependence. A person is more complex than an amoeba, the Rockefeller empire more complex than a high-street store. And both require levels of relationship between the

AN HONEST MAN BEWILDERED

Letter from Charles Darwin to Asa Gray, Downe, 22 May 1860

My dear Gray,
   Again I have to thank you for one of your very pleasant letters . . .
   With respect to the theological view of the question. This is always painful to me. I am bewildered. I had no intention to write atheistically. But I own that I cannot see as plainly as others do, and as I should wish to do, evidence of design and beneficence on all sides of us. There seems to me so much misery in the world. I cannot persuade myself that a beneficent and omnipotent God would have designedly created the *Ichneumonidae* with the express intention of their feeding within the living bodies of caterpillars, or that a cat should play with mice. Not believing this, I see no necessity in the belief that the eye was expressly designed. On the other hand, I cannot anyhow be contented to view this wonderful universe, and especially the nature of man, and to conclude that everything is the result of brute force. I am inclined to look at everything as resulting from designed laws, with the details, whether good or bad, left to the working out of what we may call chance . . . But the more I think the more bewildered I become; as indeed I have probably shown by this letter.
   Most deeply do I feel your generous kindness and interest.
   Yours sincerely and cordially,

   Charles Darwin

various component parts not called for in the simpler and more homogeneous structures. This upward, progressive movement was, he said, the fundamental law of life. There might be suffering on the way: indeed the great could only become great by struggle. Self-will and effort were the paths towards almost inevitable progress, and when applied to human society the survival of the fittest meant that advancement came from natural strength and the inherent capacity to adapt. Roughly speaking, the rich were rich because they had the prowess to become so; the poor were poor because they were lazy or incompetent, or both. Or so Spencer thought.

This concentration on self-effort was a throwback to the developmental ideas of Lamarck. He had believed that the giraffe obtained its long neck by constant straining to reach the more succulent leaves at the tops of the trees. Darwin, on the other hand, saw change as a consequence of random variations accumulated through time; by chance giraffes were born with longer necks and this conveyed a survival advantage which was passed on to their progeny. Nineteenth-century society was more attracted by effort than by luck, by advancement through industry than by winning in a lottery. Accordingly it preferred Spencer to Darwin. When Spencer went to America in 1882, the year of Darwin's death, it was like the arrival of a messiah. His books had already sold in their thousands, and everywhere he was greeted with reverence by affluent men who saw in their own wealth and selection the clear proof that society was being improved. A nation founded on self-effort, basking in new-found wealth, had found its prophet.

Within the church the doctrine of inevitable progress was seen by some as more amenable than Darwin's chance variations and scepticism over divine design. For those without the orthodox commitment who understood scientific laws as the activity of God, the theories of Darwin presented nothing but problems. They turned therefore to Spencer and Lamarck and created a gospel of progress, with God as the

Herbert Spencer used the phrase 'survival of the fittest' before Darwin. His philosophy applied the idea of evolution to a belief in social progress.

force driving evolution towards its ultimate goal. In America the eminent preacher Henry Ward Beecher wrote:

'If single acts (of creation) would evince design, how much more a vast universe, that by inherent laws gradually builded itself, and then created its own plants and animals, a universe so adjusted that it left by the way the poorest things, and steadily wrought toward more complex, ingenious, and beautiful results! Who designed this mighty machine, created matter, gave to it its laws, and impressed upon it that tendency which has brought forth the almost infinite results on the globe, and wrought them into a perfect system? Design by wholesale is grander than design by retail.'

Henry Drummond, a Scottish Free Churchman, chose in the same vein to trace the

Biologists point to the development of organisms towards greater complexity. Spencer extended this to social organization, expecting progress towards ever more complex human interactions.

development of humanity towards a nobler life. The last paragraph of one of his books reads:

> 'The Struggle may be short or long; but by all scientific analogy the result is sure. All the other Kingdoms of Nature were completed; Evolution always attains; always rounds off its work. It spent an eternity over the earth, but finished it. It struggled for millenniums to bring the Vegetable Kingdom up to the Flowering Plants, and attained. In the Animal Kingdom it never paused until the possibilities of organization were exhausted in the Mammalia. Kindled by this past, Man may surely say, "I shall arrive." The succession cannot break. The Further Evolution must go on, the Higher Kingdom come.'

Published in 1894, the book's title was *The Ascent of Man*. In 1871 Darwin had written *The Descent of Man*. There lies the difference of twenty years. Where the scientist was reticent to speak of the future, and only claimed to unlock the past, the philosopher was prepared to use evolution as a cure for the world's ills. Evolution was now spelt with a capital 'E' and looked forward rather than back. Darwin's biological theory of origins had become an all-embracing creed of progress. The cruel wastage of nature had been transmuted into a divine orchestration of design.

7
DARWIN
IN THE DOCK

On the night of 18 April 1882 Darwin confided to those who watched at his bedside that he was not afraid to die. Over the previous weeks, heart attack had followed heart attack, and recovery was now unlikely. He died the following afternoon. *The Times* acknowledged the passing of a great man whose thoughts

There was some doubt whether the agnostic Darwin should be buried in Westminster Abbey, but the service took place and the importance of his work was fully stressed.

had 'passed into the substance of facts of the universe'. *Allgemeine Zeitung* defined the nineteenth century as 'Darwin's century'.

He was buried in Westminster Abbey a few feet from the grave of Sir Isaac Newton. His pallbearers included Wallace, Huxley and Hooker. Emma had planned that he should be quietly laid to rest in the cemetery at Downe, and for some this would have been more in keeping with Darwin's own reserved nature. Others thought the Abbey unfitting for the

burial of an agnostic. But the Dean gave his approval and the service, replete with eulogies, took place on 26 April 1882.

Over the following fifty years there was always a possibility that Darwin's cherished theories would go the way of their master. The idea of evolutionary change was by then commonly accepted, but the manner in which that change occurred was still hotly disputed. Had Darwin been right to stake everything on natural selection? In later editions of *Origin* even Darwin began to doubt the total sufficiency of his new idea. Bowing to pressure, he incorporated elements of Lamarckian evolution. Then, after his death, biological science began to unlock the mechanisms of inheritance and reproduction and theory vied with theory to explain animal origins. Only in the 1920s and the years leading up to the Second World War were seemingly opposing views synthesized into a comprehensive theory. Darwin's new offspring incorporated details of how variations occurred, a problem that had long baffled Charles. But, since the new theory still echoed Darwin's focus on natural selection, it was accordingly named after him: Neo-Darwinism. We must briefly trace its birth and development.

Primroses and shellfish

During his Cambridge days Darwin was schooled in Christian apologetics, that is, the defence of religious faith against doubters and unbelievers. He rejoiced in the compelling arguments of Archdeacon Paley, that design in the natural world implied a Designer. In later years this made his own doubt the more deep as his 'design-by-natural-selection' hypothesis turned Paley's watchmaker argument on its head.

But at university world travel was only a dream. Charles had still not sailed on the *Beagle*, and it was examinations in orthodox theology that he had to navigate. In his reading he learnt the valuable lesson that discontinuities in history or in the natural world can be interpreted as evidence of God's intervention. He studied J.B.

Sumner's *Evidence of Christianity Derived from its Nature and Reception*. Sumner, then Bishop of Salisbury and later to become an Archbishop, argued that the rapid rise of Christianity within such a hostile world was evidence of God's hand at work. A purely human institution would never have accomplished so much. Darwin made notes that survive to this day. He agreed with Sumner: this sudden appearance of Christianity was proof of God's intervention.

When in later years he came to write *Origin*, Darwin was anxious to lay before the public a purely natural mechanism to account for the variety of forms in the animal kingdom. He was weary of explanation by miracle. To his way of thinking he needed to avoid any possibility of sudden changes or jumps in animal forms. Jumps implied God, and he wanted to give no open door by which miracle-mongers could enter and say it was all by acts of special creation. Each change on the evolutionary ladder must be only a very minor modification, and every movement into a new species the accrued result of innumerable small steps.

Huxley warned him at the time that this commitment to gradual change might be injurious to his theory. Might there not be some natural mechanism that allowed sudden, and large, jumps in form which were then filtered by natural selection? But for Darwin, naturalism implied 'gradualism'. To say otherwise might undermine the naturalistic tenor of his theory.

His friend's warning was prophetic. In the last decade of the nineteenth century a battle broke out between biologists who continued to support the gradualist line, and those who saw evolution progressing through jumps. The theological dimension and the question of miracles was no longer the field on which the battle was fought, but Darwin's own passionate commitment to gradual change made every scientific paper on evolution by jumps into an attack on his theory.

Ironically it started with a mistake. A Dutch botanist, Hugo De Vries, noticed some primroses in an overgrown field outside Amsterdam and found that there were differences between

SHARPENING POLITICAL AXES

George Bernard Shaw said of Darwin that 'he had the luck to please everybody who had an axe to grind'. Sharpening political weapons is always an important occupation.

We have already seen how Herbert Spencer applied his evolutionary philosophy to society as well as to nature. Big business requires free competition, and state involvement upsets the benefits which derive from the survival of the fittest. Politicians who thought along these lines were attracted to the elements in Spencer and Darwin which spoke of progress through natural selection.

But those who feared change could also quote Darwin to advantage. It may seem central to the idea of evolution that organisms (and in the social world, organizations) change. But Darwin believed the pace of change was incredibly slow. Change only occurred by 'insensible steps'. A conservative politician could therefore invoke Darwin's authority to claim that events should be left to evolve at their own speed. Policies which required government intervention or suggested radical change could be seen as going against the natural flow of history.

Then there was Karl Marx. Marx wrote to Lassalle in 1861,

'Darwin's book is very important and it suits me well that it supports the class struggle in history from the point of view of natural science. One has, of course, to put up with the crude English method of discourse.'

The central principle of Marx's work was the struggle between competing classes in society. Tensions mount until a new state of society emerges. This new society might form

naturally (for history was bound to develop in this way), or it might require a bloody revolution. Either way the nineteenth-century workers could rejoice that history was on their side in their struggle against their capitalist bosses.

At Marx's graveside his colleague Frederick Engels said, 'Just as Darwin discovered the law of development of organic nature, so Marx discovered the law of development of human history.' But Marx never made extensive use of this parallel, although the two theories were somehow seen as supporting each other. For a Marxist, Darwin emphasized struggle. And Darwin could also be interpreted as saying that development was based purely on material forces and needed no belief in God.

Finally we must mention socialism. Perhaps the key here was adaptation. Darwin talked of how organisms are adapted to their environment, and Lamarck claimed that the environment influenced future generations. The message, then, was clear. Change the environment and you will change

humanity. Darwinism implies, says Shaw in the preface to *Back to Methuselah*, that 'if we want healthy and wealthy citizens we must have healthy and wealthy towns'. Socialists accordingly advocated a programme of social change and supported those who were disadvantaged.

Survival of the fittest . . . gradualism . . . struggle . . . adaptation . . . Darwin . . . Spencer . . . Lamarck. In electing an emperor the politicians all clothed him with imaginary clothes designed, they thought, from the pages of *Origin of Species*. But the little boy who knew the truth remained at home at Downe and refused to join the political cheering.

In his oration at his friend Karl Marx's graveside in Highgate Cemetery, London, Frederick Engels compared Marx's achievement with Darwin's. Both tried to draw out laws by which development takes place — in history for Marx, in nature for Darwin. Both men have had immense continuing influence.

the plants. The tall, densely leaved stems, crowned with yellow flowers which opened towards evening (hence the name, Evening Primrose), showed considerable variation from one to the next. During the summers of 1886-88 he stayed in a nearby house and examined the varieties he found, taking plants and seedlings for his own garden. Ten years and over 50,000 primrose plants later, he had found what he believed were eight separate species. These seemed to have arisen quite by chance and not by the selective pressures proposed by the Darwinians. De Vries could not explain the origin of the changes from one flower to the next, and could only point out that the differences in species were sharply defined. He concluded that one species had 'suddenly' changed into another. But De Vries was wrong, for later research showed the differences were due to complex patterns of heredity. Nevertheless, when De Vries' book *The Mutation Theory* appeared it started an interest in evolution by sudden mutation, rather than Darwinian development through gradual selection.

Biologists quickly divided into two camps. One group, led by W.F.R. Weldon, stressed the continuity of biological forms and gradual changes within those forms. Their method was statistical, for they believed that only by examining large numbers of specimens could the slow developments in structure be traced. They were the traditional naturalists: with Darwin as their mentor they sought to trace the historical development of animal populations. To concentrate on isolated 'freaks' was to waste valuable time. Darwin had recognized the existence of sudden mutations and quoted the example of the short-legged Ancon sheep in *Origin*. But he had no explanation for their occurrence, and assumed they were quite irrelevant to the process of evolutionary development.

The opposing school of thought pinned their hopes on these very mutations. Here, they believed, was the mechanism by which nature jumped up the evolutionary ladder. William Bateson did research on organisms in the Aral Sea. The sea was becoming increasingly salty and Bateson was fortunate enough to find a series of lakes which differed from each other in salinity. Darwinism predicted that, since an organism was adapted to its environment, there would be a smooth scale of differences as the salt level increased. In fact Bateson observed that the shellfish had marked differences between one lake and the next. Such changes could not, he argued, be the result of the environment. They owed their origin to hereditary mutations.

In the previous chapter we have been cautious in using military idioms to express disagreements over ways of thinking. Too readily the mind boxes complex issues into opposing 'camps'. And yet between Weldon and Bateson the analogy is fair. The two men were bitter opponents, seldom conceding an iota of truth in the other's position. The science journals at the turn of the century rang with charge and counter-charge, research paper and fervent refutation. The clash of personalities did much to damage real progress. One side dismissed mutations as harmful freaks and a diversion from the real study of gradual evolutionary change. The other group acknowledged the small-scale variations of which Darwin had made so much, but saw them as fluctuations about a norm. Such variations could never add up to a complete change in species. Nature needed sudden, and large, jumps to evolve.

In the end the two views would be synthesized, with both theories playing their part. But before that was possible some deeper understanding had to be gained on how the hereditary process worked. It was essential to unlock the source of these mutations (if they existed) and to trace how one organism passed on its selective advantage to another.

Pea plants and the death of Darwinism

Though a devout monk, Gregor Johann Mendel was no recluse. We have already seen how he travelled to London to see the 1862 Exhibition,

GREGOR MENDEL

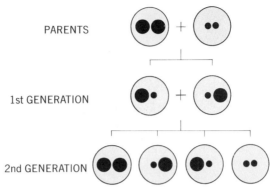

YELLOW FACTOR: DOMINANT. If present pea is yellow.

GREEN FACTOR: RECESSIVE. Always over-ridden if yellow factor is present.

PARENTS

1st GENERATION

2nd GENERATION

Johann Mendel (1822-84) was born in Austrian Silesia, the son of poor peasants. He studied at the University of Vienna under some of the leading physicists and biologists of his day, including the famous Professor Doppler of 'Doppler Effect' fame, who explained the change in sound pitch as an object passes at speed.

The young Mendel was ordained (adding the name Gregor) and entered a monastery where, from 1856, he conducted his experiments with generations of pea plants. He isolated plants with particular characteristics, interbred them, and traced the ebb and flow of the characteristics from one generation to the next. Along with five other traits, he looked to see if the peas produced were yellow or green, and whether plant stems were long (around 2 metres) or short (0.5 metres).

In one experiment he cross-bred pure strains of yellow peas with green ones but found that the first generation of offspring were all yellow. Green peas seemed to have disappeared. But then when these yellow peas were in turn crossed the 'lost' green factor returned. From 258 pea plants Mendal obtained 8,000 peas of which 6,000 were yellow and 2,000 green. Any notion of 'blending inheritance' would expect peas that were a mixture of yellow and green, or even yellow with green spots! It seemed that the inherited colour was either one hue or the other and could be suppressed. From this Mendel deduced that inheritance passed on particular characteristics, which remained intact in the offspring and were not blended.

But why, he asked, were 6,000 of the plants one colour and only 2,000 the other — a ratio of three to one? That this was no coincidence was shown by other experiments: long stems always dominated by the 3:1 ratio as well. Mendel concluded that each parent in reproduction gave one colour characteristic to its offspring. Whether this was a yellow factor or a green factor was a matter of chance. So, in the 8,000 peas, 2,000 had been given two green factors, 2,000 had been given two yellow factors, 2,000 a yellow and a green factor, and 2,000 a green and a yellow factor. Mendel could explain the 3:1 ratio if the yellow colouring was 'dominant' in that it overrode any tendency to produce a green seed. Green would only arise where there was no yellow factor at all. And that was only in the first batch of 2,000 peas that had received two green factors.

Mendel, of course, could not explain his findings in terms of modern genetics, which sees the factors as specific genes within the reproductive cells. Nevertheless, he had stumbled on clear evidence that inheritance involved the passing of discrete characteristics, and that certain of these characteristics dominated over others. Particular traits could therefore be suppressed for one or more genera-

This diagram shows two generations of pea plants. In the second generation, as Mendel found, the dominant yellow peas outnumber green peas by three to one.

tions, only to appear much later.

He presented his findings to the Natural History Society of Bonn. The *Proceedings* carried the research paper in 1866, and copies were automatically dispatched to over 100 other learned societies, including the Royal Society and the Linnaean Society in Great Britain. Mendel also sent an offprint to Nägeli, a leading botanist. Unfortunately Nägeli was fully committed to the blending view of inheritance, and simply dismissed Mendel's experiments. In his important book of 1884 on evolution and inheritance, he makes no mention of Mendel — another instance of a prior viewpoint acting like a pair of blinkers.

In 1868 Mendel was made prelate of his foundation, and a few years later the monastery became embroiled in a tax dispute. There was no more time for research. In 1884 he died, known to naturalists only by a fuchsia named after him for his services in adjudicating flower shows.

and he widely circulated to other scientists his important paper on hybridization in peas. He was not unknown, only unappreciated. By some he was dismissed as an amateur; a great pity, for he held the first keys to inheritance.

By crossing thousands of peas of different characteristics, Mendel was able to show that heredity worked by passing on certain characteristics or factors from one generation to another. In an offspring there was no 'half-and-half' blending of the parents' characteristics, like two coloured paints, as Darwin had supposed. The characteristics were passed on intact, although certain factors seemed to take preference over others and were accordingly called 'dominant'. Other factors were 'recessive', and could lie dormant for many generations before revealing themselves again. For us it is common to notice a particular characteristic we have inherited from our grandparents, which was not present in our parents.

Mendel traced the path of these characteristics as they passed from one generation to the next, and worked out exact mathematical ratios showing how much of each characteristic appeared in any generation. Of the mechanism involved Mendel had no idea — or if he did he told no one. The crucial point was that inheritance occurred by passing on particular factors, not by simply mixing parental characteristics. Here was the answer to Fleeming Jenkin who, to Darwin's embarrassment, had demonstrated that any useful variation would be immediately swamped through subsequent breeding. On Mendel's scheme an evolutionary change was inherited intact and, if dominant, would override the characteristics of the more mundane parent. But Mendel had little interest in the origin of species, and Darwin no grasp of mathematics. The two men never met or corresponded, and Darwin was left with his puzzle.

With the new century, the interest in biological research had moved. Now the question of heredity was central, and the past literature was scrutinized for work which might illumine the subject. De Vries came across a reference to Mendel's paper and quickly saw the relevance

Gregor Mendel's botanical researches, conducted in the gardens of his monastery, opened up an understanding of genetic inheritance.

of his work. Other researchers repeated the experiment with different plants and even with mice. Mendel's results were confirmed and the monk, sadly now long dead, became famous.

Mendel's discovery tended to add weight to the 'jump' theories for evolution. It showed how a sudden mutation might propagate down the generations. Bateson dismissed Charles Darwin's earlier theories:

> 'We read his scheme of Evolution as we would those of Lucretius or of Lamarck. The transformation of masses of populations by imperceptible steps guided by selection is, as most of us now see, so inapplicable to the fact that we can only marvel both at the want of penetration displayed by the advocates of such a proposition, and at the forensic skill by which it was made to appear acceptable even for a time.'

But the new Mendelism was still only a mathematical analysis of the flow of inheritance; it gave no clues as to its mechanism. The factors transmitted from one generation to the next were dubbed 'genes', but initially many biologists doubted whether they were anything more than working concepts. No actual 'particles' could be isolated.

Furthermore, the idea of inheritance by blending was still not dead. Analysis of human characteristics, such as height, showed continuous and gradual variation across generations of families. If Mendel's theory was true, then one would expect a number of particular heights, as each generation picked up the transmitted 'height factor' from its forebears. Certain plants and animals clearly show blending inheritance. Or so it seems. The petals of snapdragons reflect a mixture of the previous generation's colouring, and crossing a white-haired American shorthorn cow with a red-haired bull produces calves with both red and white hairs.

But even the opponents of Mendelism were no friends of Darwin. True, they stressed gradual change and a continuous spectrum of variations. But they no longer saw natural selection as the all-sufficient mechanism for evolutionary advance. Either they believed there was some mysterious evolutionary pressure driving animal adaptations along a particular pathway, or they revived the ideas of Lamarck that changes during the lifetime of an individual animal could be passed on to the offspring. It mattered little that August Weismann, professor of zoology at Freiburg and former physician to the Archduke Stephan, had failed to find evidence for this kind of 'soft inheritance'. He had cut off the tails of one-and-a-half thousand mice, but in over twenty generations not one mouse had been born tail-less. Where, then, was the hoped-for inheritance of acquired characteristics? For the professor it was clear that the old story of a blacksmith passing on strength to his son was a fable. Changes in the cells in one part of the body (say, in strengthening the arm muscles) did not correspondingly affect those cells which produced the eggs and sperm to form the next generation.

A combination of intrinsic evolutionary drives and Lamarckian inheritance seemed to make better sense of the data. During the first quarter of the twentieth century Darwin's theories were all but dead. But as the old Darwinism seemed to drift into the sleep of death, there were many in the United States who feared that the same fate could befall Christianity. Secular ideas and values were suffocating traditional faith.

Mythology, mistakes and meaning

The impact of the historical approach had caused as much of a revolution in theology as in biology. And to understand the meaning of some of the twentieth-century conflicts over evolution we need to look at how views developed about the nature of biblical teaching. We have seen how *Essays and Reviews* (1860) mimicked Darwin's *Origin of Species* in seeking out natural causes rather than supernatural intervention. God may still speak to us through the Bible, but the historian was bound to treat the Scriptures 'as any other book'. But once the status of the Bible had been undermined Christianity was stripped bare of any power to influence or change mankind. Or so it seemed to many. And there was a widespread attempt to return to Christian orthodoxy from the liberalism of the previous decades.

August Weismann's genetic experiments with mice proved that characteristics acquired during an individual's lifetime cannot be passed on.

For the theologians the nineteenth century created two new approaches to faith.

☐ *First, it opened the door to the historical study of the Bible documents.* Where did they come from? Were the early books all written by Moses, or all the New Testament letters by the apostle Paul? What was the context which led to the selection or rejection of material which gave us the Bible as we know it? How did the faith of the Hebrew people develop, or evolve?

As with Darwinism this procedure was not, of necessity, threatening to faith. In the hands of the Cambridge scholars Westcott, Lightfoot and Hort it established faith on a firmer footing by uncovering the trustworthiness of the ancient records. Others, of different persuasion, used historical criticism to bolster their own philosophy. One such was F.C. Baur in Germany, a disciple of Hegel, who tried to fit the history of the early church into a pattern of conflict between the apostles Peter and Paul leading to resolution in the second century. Others tried to find within the Old Testament a picture of

In 1835 David Strauss published his 'Life of Jesus Critically Examined'. He adopted a radical approach to the Bible's historical claims.

religion evolving from crude polytheism to the noble idea of a single God. Historical criticism was a valuable tool; how one employed it depended on prior suppositions.

☐ *Second, as this critical enquiry began to question the historicity of the Bible, theologians tried to distance the beliefs of Christianity from its historical locus in the Palestine of the first century.* Christianity is embarrassingly historical. Christians believe that at a certain time and place God lived the life of a man. The validation of this lies in the resurrection of Jesus, an event which can be dated to within a few years and which the early disciples took as foundational for their faith. As Paul, an early convert, expressed it, 'If Christ has not been raised, then your faith is a delusion.'

With the nineteenth-century vogue to discover only natural explanations, belief in the resurrection was indeed embarrassing. So was the apostle Paul deluded? A young German scholar, David Strauss, came up with a new, and somewhat radical, solution.

In 1825 Strauss entered the university of Tübingen as a student. By 1832 he was an assistant lecturer, and three years later he published his *Life of Jesus Critically Examined*. The book had a tremendous success, making him famous yet destroying his career. When he was offered a university chair at Zurich in 1839, a 40,000-signature petition against him caused the Swiss Government to prevent him taking up the appointment. But in writing his *Life of Jesus* Strauss was trying as a sincere Christian to steer a middle way between those who dismissed the miracle stories as impossible and those who tried to find purely natural explanations, such as that Jesus was really walking on logs when the disciples believed he was walking on water.

Within these two schools the argument was 'did the miracles actually happen?' It was this very question that Strauss wanted to avoid. Instead of debating whether the miracles had actually happened or had only appeared to happen, he concentrated on the significance the events held for the disciples. He claimed that the miracle stories expressed not history but the way the disciples revered Jesus. They tell us more about the disciples' beliefs than the actual events.

Take the feeding of the 5,000. Atheists simply dismissed the story as absurd, for things like that just do not happen. Rationalists retrieved the story by saying that Jesus inspired the crowd to share their lunch packets. For Strauss the truth lay elsewhere. The disciples believed that Jesus was the new Moses, the prophet long foretold in the Old Testament. Moses had fed the people of Israel with manna as they wandered through the wilderness. Since Jesus was a prophet greater than Moses, then he must have fed the crowds in the same way. The story was not wilful invention, a lie concocted by the disciples for propaganda purposes. It had arisen quite naturally out of a community deeply committed to Jesus as Messiah. It reflected, not history, but the living significance of Jesus. But the story was still 'true', in the sense that it accurately expressed the disciples' belief.

In the thinking of Strauss the age-old debate over miracles was circumvented. It mattered little what had actually happened. More important was the faith created and the theological truths expressed. In studying the New Testament the focus of interest had moved from the question 'What actually happened?' to 'What significance did it hold for the disciples?' And even more important, said the new outlook, is the faith these ancient stories evoke today. In stripping from Jesus all the fanciful miracles we are not left with nothing. We find a deeply inspired prophet from Nazareth who imparted a new dimension of religion and morality to his generation — and now to ours. He so impressed his hearers with the wonder of God that they could only express these new insights in tales of miracles. It was the only vocabulary they had. The task of the church was to enter into this religious depth, and to focus our wills to serve others as Jesus had done.

For many orthodox Christians all this was very worrying indeed. The Bible, as a historical revelation of God's acts among men and women,

was being turned into a picture book to illumine a rather general belief in the love of God and the brotherhood of all mankind. Jesus was not really a divine Saviour, but a man like us who had somehow broken through to new depths of religious insight. We might learn from him, but what had happened in the past was of little consequence for our present faith.

Genesis and the creation stories of Babylon

While Strauss threatened the New Testament, archaeologists unearthed questions for the Old. Excavations in Babylonia revealed the palaces of ancient kings, some of whom were contemporary with Isaiah and the fall of Jerusalem described by the biblical prophets. In Victorian England each new discovery was met with

This ancient Babylonian tablet shows a plan of the world known at the time. Babylonian stories of the world's creation were in existence before the Genesis account was written down.

fascination, and proved the existence of ancient rulers otherwise known to us only through the pages of Scripture. But among the broken clay tablets, covered in cuneiform writing, were some with references to 'creation'.

Like a jigsaw, the pieces were assembled to reveal an account of the creation of the world. Today the difference between Genesis and the Babylonian account is evident. The first speaks of one God creating the world and mankind by his own command; the other describes chaos and war among many gods, after which one god, Marduk, fashions humanity from clay and blood. The spiritual depth and dignity of Genesis far surpasses the polytheistic ideas of Babylon. Yet until the complete story had been reconstructed, incautious scholars talked of the Bible account being a copy of that from Babylonia. Certainly, they argued, Genesis should be consigned to the category of legend, and its writing was dated long after Moses to the time Israel was held captive in Babylon.

Much of nineteenth-century liberalism has now been shown as excessive. The Old Testament is not a poor reflection of more ancient Babylonian or Canaanite tales. There are more differences than similarities between the texts. The opening chapters of Genesis stand unique. Nevertheless, many scholars still use the category of myth in relation to some of the biblical material. Are they right to do so?

We glibly assume that the word 'myth' is equivalent to a fictitious tale. But when scholars call creation stories myths they do not mean that they are sweet fables suitable only for children. Ancient world literature abounds in creation stories. They are attempts to convey the meaning behind a people's existence. They pinpoint what is central to life, be it the sovereignty of the gods or the sacredness of the fruit-bearing earth. They may be cast as historical stories, beginning with the first moments of time, but their lessons are not about chronology. They are poetic tracts on humanity's relationship with the gods and with creation. They are not so much pre-scientific (as though a simple people could not understand the origin of their

world), as 'trans-scientific'. They are a unique genre of literature that expresses fundamental truths which reach way beyond the limits of scientific terminology. They are deeper than scientific description, as poetry is deeper than a local government circular.

On this understanding even a twentieth-century description of cosmic origins can take on a mythical dimension. Carl Sagan's writing in *Cosmos* (1980) is a case in point. The words are apparently scientific, yet his intention is to overawe us with the majesty of creation. He wants to satisfy our deep longings for an understanding of our origins, and to impress upon us what heights of consciousness and creativity we have reached without the help of gods. Theodore Zeldin, writing on French society in the 1980s, describes Sagan as 'an example of an intellectual who has turned science into poetry for mass consumption . . . His manner is neither French nor American but simply nineteenth century. There is still a demand for that.'

Sagan himself believes his picture of creation 'has the sound of epic myth, and rightly'. And from the Genesis account he deliberately borrows words such as form, void, darkness and light to evoke echoes in our mind. Like Genesis, he starts with a cosmos in chaos and ends with the creation of humanity. The language may reflect scientific facts and the sequence of creation, but the story has a different purpose. Sagan is trying to assure our souls, and awaken our commitment. Above all he is trying to give mankind dignity 'by the courage of his questions and the depth of his answers'. His account of creation is the story of nature's growing self-awareness, and it is interesting that (like the biblical account) he treats mankind as the pinnacle. Sagan himself wants to create faith, not impart science.

It is not a faith, of course, which the Bible's authors would relish. It may give glory to mankind, but it offers not a whisper that we are created after God's image. And this is what Genesis wants to shout out. We may toy with many 'creation stories' to satisfy our longings, but Genesis wants to put the record straight. We were created by God.

With this in mind the early chapters of Genesis may be seen as an antidote to false faith rather than as a lesson in biological origins. Within the structure of the Bible's first chapter there is much to suggest a stylistic form. The words may not be cast in the rhythms of Hebrew poetry, but that does not exclude the possibility of a literary rather than a scientific structure. We must find the clues within the text. And here we can note a parallelism between the first three days of creation and the last three days. In days one to three, God creates the light, the sky and water, and then the earth. In days four to six he makes occupants for these realms of creation: the stars, the birds and fishes, the animals and people. The message is not scientific, but theological. The author is tracing out the totality of creation. We owe our being to God. The whole world is his, with nothing outside his power.

As Sagan has borrowed from Genesis, so the style, and even vocabulary, of this beautiful first chapter of the Bible may have been borrowed from Babylon. If so, the borrowing was deliberate, with the author using the thought forms of his day to express yet deeper (and older) revealed truths about God. Where the Babylonians believed in many gods, Genesis stresses that God is one. Where the Babylonians sought to appease their rather capricious deities, Genesis upholds God as benevolent and providential. Where the Babylonians had to bolster faith in an orderly existence by an annual recitation of their story, Genesis simply says that God now rests, having made everything good. If the writing of Genesis can be dated to the time the Jewish people were in exile, then what better instrument to instruct the young, brought up in a culture riddled with false beliefs in many gods?

In this light Genesis is *anti*-mythological; it seeks to correct the false ideas of influential Babylon. It seeks to restore true faith.

The rise of fundamentalism

For liberals such as Oliver Wendell Holmes in Boston, the conclusion was clear:

COSMOS — THE STORY OF COSMIC EVOLUTION

(CARL SAGAN, 1980)

'For unknown ages after the explosive outpouring of matter and energy of the Big Bang, the Cosmos was without form. Deep, impenetrable darkness was everywhere, hydrogen atoms in the void . . .

'A first generation of stars was born, flooding the Cosmos with light. There were in those times not yet any planets to receive the light, no living creatures to admire the radiance of the heavens . . .

'In the dark lush clouds between the stars, new raindrops made of many elements were forming, later generations of stars being born. Nearby, smaller raindrops grew, bodies far too little to ignite the nuclear fire, droplets in the interstellar mist on their way to form the planets. Among them was a small world of stone and iron, the early Earth . . .

'Congealing and warming, the Earth released the methane, ammonia, water and hydrogen gases that had been trapped within, forming the primitive atmosphere and the first oceans . . .

'Molecules were organized, and complex chemical reactions driven, on the surface of clays. And one day a molecule arose that quite by accident was able to make crude copies of itself out of the other molecules . . .

'Single-celled plants arose, and life began to generate its own food. Sex was invented. One-celled organisms evolved into multicellular colonies, elaborating their various parts into specialized colonies. Eyes and ears evolved, and now the Cosmos could see and hear . . .

'Small creatures emerged . . . They survived by swiftness and cunning. And then, only a moment ago, some small arboreal animals scampered down from the trees. They became upright and taught themselves the use of tools, domesticated other animals, plants and fire, and devised language. The ash of stellar alchemy was now emerging into consciousness. At an ever-accelerating pace, it invented writing, cities, art and science, and sent spaceships to the planets and the stars.

'These are some of the things that hydrogen atoms do, given fifteen billion years of cosmic evolution.'

'The truth is staring the Christian world in the face, that the stories of the old Hebrew books cannot be taken as literal statements of fact.'

But many on both sides of the Atlantic did not like to be stared at. Especially when the face of the so-called 'higher criticism' was not seen as truth but as the shadowy visage of German professors and sceptics. As early as 1879 a professor in the Southern Baptist Theological Seminary was forced to resign following complaints about his teaching; he appeared to doubt the full inspiration of the Bible. 'Our denomination will arm itself against this crusade of vainglorious scholarship,' thundered one Baptist newspaper. 'The fortunes of the kingdom of Jesus Christ are not dependent upon German-born vagaries.'

There was a backlash against theological liberalism, and a movement arose which stressed the literal and straightforward meaning of the sacred text. In America at the turn of the century the Presbyterian General Assembly drew up a list of fundamental beliefs. These were seen as the bedrock of Christian faith and they included a statement on the inerrancy of the Bible. All this linked well with an American emphasis on self-evident knowledge and 'common sense'. The Bible was to be taken at its face value, and the accounts of miracles (be they in the Gospels or in Genesis) taken as simple history. The movement was not, at its outset, against theories of evolution. In their journal, *The Fundamentals*, appeared one or two articles from an evolutionary perspective. They spoke of the tradition of science discovering the pathways of God. But as the movement quickened, the emphasis on the accuracy of the Bible text was quickly transmuted into an insistence on its literalism. What the Bible said, it said. Simple as that.

As for Strauss and his mythological interpretations, some linked the collapse of Germany in the First World War to corrupt biblical scholarship. It was 'infidel Germany against the believing world — Kultur against Christi-anity — the Gospel of Hate against the Gospel of Love.'

There was more to the rise of the fundamentalist movement than scholarly disagreement over the way the Bible should be read. It was also a reaction against changing standards of morality, and fear of alien political movements such as communism. War and subsequent demobilization, automobile and tractor, radio and mass education, all accelerated change. Following the First World War the farms in the south of the United States changed over from traditional cotton to tilled crops and livestock. Family-based farms were replaced with large commercial combines. The beginning of the century saw an increasingly industrialized society sweeping away the older, rural, America. In the face of massive change people cling to their past. The 'Back to the Bible' campaign meant a restoration of the securities of the past.

But to trace fundamentalism's sociological roots is not to explain it away. The renewed emphasis on taking the Bible literally was not just a product of political fear and social instability. At heart was a profound division in religious belief:

☐ *The liberals discarded the Bible as a direct revelation of God, and sought God's voice in the inner conscience.* The Bible, they said, testified to the workings of God in the Hebrew people down through the ages, and supremely in Jesus. But, the liberals said, it was only the story of those events, and the memoirs of those who experienced them. The Bible words themselves were all too human recollections.

☐ *The fundamentalists disagreed. The words of both Old and New Testament, although penned by human authors, were the very words of God.* The Bible was to be understood literally, from Genesis to Revelation. To admit error in one part was to destroy the whole. Rather, in exact statements of truth, the Bible revealed the nature of God and his plan for mankind.

Many, on both sides of the Atlantic, refused

to accept that the choice lay between such stark contrasts. For them the Bible remained the trustworthy revelation of God. Yet they recognized the humanity of the Bible writers and pursued any historical research that would clarify the meaning of the text. They tried to unlock the past by studying the culture of Babylon, Jerusalem and Rome. God had spoken in the Scriptures, but just as Jesus wore the clothes of a first-century carpenter so the texts were expressed in the idioms of their day.

Between the ends of the spectrum it was inevitable there would be another clash over evolution and Genesis — a clash between a society fashioned by liberal progress, and those whose way of life derived from unchanging standards revealed in an inerrant Bible.

This time it was a 'battle'. Science had little to do with it.

The Scopes Trial

From the end of the Great War, fundamentalism gathered both momentum and militancy. During the early years of the 1920s the movement gained cohesion, and drew up two distinct battle arenas. In the first, their aim was to combat the spread of religious liberalism among the church denominations. In the second, they sought to prevent the teaching of evolution in the public schools. The history of the struggles to control church boards and conferences need not detain us. Concerning evolution, the fundamentalists were on the brink of achieving major legislative successes, when the movement seemed to dissolve about them. Instead, they were engulfed in public ridicule. The date was the summer of 1925. The place Dayton, Tennessee.

John Scopes, a high-school science teacher, was accused of teaching evolution in defiance of a recently enacted state law. Other states had considered anti-evolution laws, and Oklahoma had passed an amendment to a textbook bill which forbade any 'materialistic conception of History (i.e.) The Darwinian Theory of Creation vs. the Bible Account of Creation.'

That Charles Darwin had assiduously avoided the matter of 'creation', and had limited himself to understanding the derivation of species, was lost on them. There were bigger issues at stake. As one proponent explained:

> 'There is a fundamental difference between letting a person think as he pleases and in using time and service which the state pays for and which belongs to the state in teaching children of others theories that are not Biblical.'

Attendance laws now insisted that parents send their children to school. In most cases this meant a state school. Parents were worried that a state education would lead their own children away from cherished family beliefs and values. In 1925 this parental concern gained its first substantial legislative success. Three years earlier in Kentucky they had been voted down, but in Tennessee the bill to ban 'any theory that denies the story of Divine Creation of man as taught in the Bible' was accepted seventy-one to five. And it was against this newly passed bill that Scopes had offended.

Scopes was clearly guilty, and was accordingly fined $100 — a fine later overturned on a technicality of who had set it. But the money and the verdict were unimportant. On trial was Darwin's theory of evolution and the freedom to teach such theories in the classroom.

As prosecution lawyer, opposing Scopes and evolution, stood William Jennings Bryan, presidential nominee of the Democratic party and one-time Secretary of State under Woodrow Wilson. Throughout his public life he had spoken out for the underprivileged and busied himself in international concerns for peace. He was a man of national stature and fully committed to basing his vision of society on the Bible. For him, evolution spelt destruction of the Bible and the moral society founded upon it.

> 'All the ills from which America suffers can be traced back to the teachings of evolution. It would be better to destroy every book ever written, and save just the first three verses of Genesis . . .'

During the trial he spelt out what he meant:

'Why, my friend, if they believe (in evolution) they go back to scoff at the religion of their parents. And the parents have a right to say that no teacher paid by their money shall rob their children of faith in God and send them back to their homes sceptical, infidels, or agnostics, or atheists.'

The defence employed as their lawyers the fabled criminal attorney Clarence Darrow, noted as an agnostic, and W.R. Malone, who had formerly served under Bryan at the State Department. The judge reluctantly allowed the case to extend beyond the simple limits of whether or not Scopes had broken the state law, into a debate on the rights and wrongs of teaching evolution. Dayton itself enjoyed the spectacle, with all the earnestness of a religious war yet the revelry of a party. The town was awash with banners: 'Where will you spend eternity?' The stores sold pins reading 'Your Old Man's a Monkey', and circus performers arrived with chimpanzees.

Sadly Bryan, a first-rate politician but no Bible scholar, allowed himself to be cross-examined as an 'expert witness' on the Bible. With Bryan in the witness box, the sharp-witted Darrow ruthlessly highlighted his opponent's ignorance, and mocked his inconsistencies. At the end the state law was upheld; the high-school teacher had broken the law. But the lasting image, at least with the intelligentsia back in New York, was of pitiful ignorance and superstition arrayed against science and learning.

In allowing himself to be so questioned Bryan may have made a mistake, but he also played true to the anti-evolutionary cause. The fundamentalists were not against science. Far from it. They supported the teaching of science in schools. They were, however, committed to a philosophy which believed all truth to be self-evident and clear to any person of common sense. Three years before, Bryan had

supported a Kentucky effort to stop the teaching of evolution in their public schools:

'Commit your cause to the people. Forget, if need be, the highbrows both in the political and college world, and carry this cause to the people.'

As he testified before the people of Dayton, Bryan therefore believed he had no need of expert witnesses, be they biologists or theologians. It was patently obvious that mankind was different from the monkeys. And against such clear 'facts' were marshalled the mere suppositions of the Darwinists. In his eyes it was hardly a contest. Again and again the anti-evolutionists emphasized that science rested on facts; evolution was a mere hypothesis. As Bryan wrote in a speech intended as the lynch-pin of the trial:

'It is not scientific truth to which Christians object, for true science is classified knowledge and nothing can be scientific unless it is true. Evolution, on the other hand, is not truth; it is merely hypothesis — it is millions of guesses strung together.'

Bryan was not prepared to 'guess' when the Bible clearly told him the truth. And he was not prepared to see state education used as a tool to promote evolutionary guesses, to the detriment of traditional morality and values.

There was a sad end to the trial. Of course the anti-evolutionists won their case. For, regardless of the wisdom of ever enforcing the 'monkey law', Scopes had certainly broken it. But five days afterwards Bryan died. Dayton founded a small college in his name, which remained a stronghold of anti-evolutionism.

The anti-evolution law was not repealed until forty years later. However, in other states similar laws now proved impossible to get through, and only those in Mississippi and Arkansas were passed. There was increased caution on behalf of textbook publishers and on the appointment of teachers. In 1924 North Carolina's board of education announced that it

THE CHICAGO DAILY NEWS.
BOX SCORE

LL EDITION   TUESDAY, JULY 21, 1925.   HOME EDITION   5 O'CLOCK

50TH YEAR—173.

SCOPES "GUILTY" IN APE CASE

John Scopes

William Jennings Bryan was the prosecuting lawyer at the Scopes trial.

The Scopes trial, in Dayton, Tennessee, ranged beyond the question of what the defendant had taught. The rights and wrongs of teaching evolution were debated at length.

LATE BASEBALL SCORES.

Hawthorne Race Results.

DEMANDS BIG INQUIRY INTO SLOAN'S PAVING

FIGHT ON TO SAVE SCOTT

Detroit Clubwomen and Newspaper There Start a Campaign.

SCOPES FOUND GUILTY, FINE OF $100 IMPOSED

Judge Allows 30 Days for Appeal; Trial Winds Up with a Rush.

JURY OUT NINE MINUTES

BY HERBERT M. DAVIDSON

(Special Dispatch from a Staff Correspondent.)

Dayton, Tenn., July 21.—A quick verdict was returned in the Scopes trial today.

WINISM IN TENNESSEE.

THE TENNESSEE TRIAL.

CASE FOR DEFENCE.

(FROM OUR OWN CORRESPONDENT.)

NEW YORK, July 14.

After a comparatively dull morning's argument in regard to the constitutionality of the anti-evolution law, Mr. ...

THE TENNESSEE TRIAL

SCIENTIFIC EVIDENCE FOR DEFENCE.

(FROM OUR OWN CORRESPONDENT.)

NEW YORK, July 16.

The case for the State of Tennessee in the prosecution of Mr. Scopes was completed in less than two hours yesterday afternoon.

CLEVELAND PLAIN DEALER

ALL OHIO EDITION

26 PAGES   CLEVELAND, WEDNESDAY MORNING, JULY 22, 1925   PRICE TWO CENTS

COPES GUILTY, WILL FIGHT ON

5,000 CHILDREN IN PARADE AS AKRON CENTENNIAL EVENT

TEACHER IS FINED $100; FLAYS LAW

Shirt-Sleeved Defendant Tells Court Statute is Unjust; Counsel to Appeal Case.

JULY 18, 1925

THE TENNESSEE TRIAL.

SCIENTIFIC EVIDENCE DISALLOWED.

OPPOSING SPEECHES.

(FROM OUR OWN CORRESPONDENT.)

NEW YORK, July 17.

would not adopt biological texts which contradicted Genesis. Some publishers complied by removing offending sections. And in other states the religious orthodoxy of any would-be teacher was tested before he or she was appointed. But still full-blown anti-evolutionists did not carry the day in America's schools.

In the nineteenth century the Oxford clash between Thomas Huxley and Bishop Wilberforce gradually took on — at least in the eyes of Darwin's friends — the status of a myth symbolizing their struggles between the old guard and the new. In contrast, the well-publicized Scopes trial pinpointed a very real divergence of social and religious views. The defence was not anti-religion. The defence lawyer Malone was himself a Catholic, and remained so after the trial. The testimony of his colleague Darrow had not shaken his faith, for it was perfectly possible to see evolution as the method of God's working in the world. Nor indeed was the scientific discipline itself questioned by the Christian fundamentalists. They sincerely believed in the pursuit of scientific truth. On trial in Dayton was neither religion nor science.

So what was the argument? The fires of dissension were three, set alight by the sparks of two clashing cultures, the urban and the small-town rural:

☐ *The method of doing science.* Where fundamentalists demanded proven facts and demonstration, the evolutionists were content to accept circumstantial evidence to piece the jigsaw of nature together. To them a theory was acceptable if it linked together and explained the chaos of observed phenomena.

☐ *The status of the Bible.* The fundamentalists wanted to interpret Scripture literally and to regard Genesis as a straightforward account of creation. Sudden interventions by God were part of his creative activity. The evolutionists sought out a sequence of natural cause and effect, and doubted the Bible's relevance to their search.

☐ *Freedom of intellectual enquiry.* Bryan claimed that evolution in the classroom destroyed traditional values. Darrow feared that restrictions in one educational subject might lead to restrictions in others. To allow the anti-evolution law to stand was to open the doors to further restrictions or manipulation of the teaching syllabus. Not all such pressure would come from God-fearing Christians, and it was a path back to the bigotry of the sixteenth century.

All these we have met before. Darwin himself was criticized by his fellow naturalists for spinning a web of unprovable hypotheses. Lyell tried to establish a science based on the gradual accumulation of natural causes. Huxley feared that bibliolatry would hinder scientific research by posting warning signs, 'No thoroughfare. By order. Moses.' The setting in Dayton was far removed from fog-bound London or sleepy Downe. By 1925 the science of animal origins had progressed way beyond Charles' 'theory to work by' of 1842. The social milieu that set country farmer against city dweller was totally different from the struggle for dominance in Victorian England. Yet the questions were the same: the methodology of science, the status of the Bible, and the freedom to enquire.

The new synthesis

Away from the courtroom, research continued on unravelling the mechanisms of evolution. The search was on to find the particles, or genes, which held the hereditary information. Work with cell structures provided the next step.

For some years the existence had been observed of tiny threads within every organic cell. Each cell in any particular organism had the same number of these threads joined in pairs, although the number varied between as few as two pairs to more than 650 pairs (both extremes being found in plant cells), with human cells containing 23 pairs. The threads readily picked up the colour of the dye used to stain the cells for viewing under a microscope, and accordingly they became known as 'chromosomes', from the Greek word for colour. As a cell divided so these threadlike objects first doubled in number and then divided equally between the two cells. During the first decade of this century it was shown that it was these structures that transmitted the hereditary information. More exactly, they appeared to be the carriers or messengers for the particular characteristics locked up in the genes.

T.H. Morgan spent years breeding *Drosophilia melanogaster,* known to most of us as fruitflies. His aim was not to repeat Mendel's work, but to see if mutations occurred and to identify the chromosome carrying the gene responsible. It was as though the genes (whose very existence Morgan at first questioned!) were beads strung along the chromosome thread. If certain genes were not transmitted to the next generation then the offspring might be different.

In the 1920s H.J. Muller bombarded cells with X-rays and found that the genes were often altered leading to new organic structures. Evolutionary change was believed to arise from ordinary recombination of genes (as Mendel had described) with additional possibilities through the alteration of the genes themselves. But again the biologists were brought up short for lack of knowledge. As a cell divided, how was the transfer process accomplished? What caused the changes in genes between one organism and its neighbour? They did not know. But clearly all this was a long way from Darwin and natural selection. Gone was talk of nature 'red in tooth and claw'. It seemed that evolution was at the mercy of microscopic particles, called genes, whose behaviour was unknown.

The bringing together of the new ideas in genetics, with the old Darwinian theory of evolution by natural selection, was initially the work of one man, Ronald Fisher. Fisher, who had studied statistical mechanics at Cambridge University, came up with a mathematical model which showed that the 'gradual' variations observed in nature could indeed be the product of 'bitty' Mendelian inheritance. Given enough Mendelian factors, or genes, the net effect of generations of recombination would be a

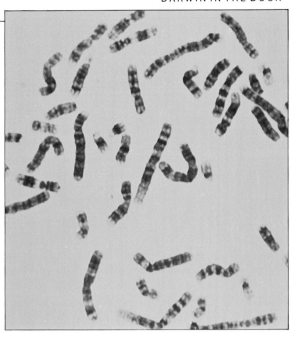

This colour-enhanced micrograph shows male chromosomes during cell division.

gradual spectrum of forms. A television picture is a combination of specific dots of three basic colours, yet with sufficient dots their combination can produce delicate shading and all the colours of the rainbow.

Mendel had been lucky in selecting a plant in which each characteristic was controlled by just one gene. It is more common, though, for one gene to influence a number of characteristics, or one characteristic to be controlled by more than one gene. Fisher stressed that in most instances changes in genes had only minute effects on the shape and structure of the grown adult form. Gene mutations, or the new genetic structure arising through reshuffling in reproduction, provided only small alterations. Large jumps were unusual. Here, Fisher claimed, were the small and infinitely varied changes between one generation and the next of which Darwin spoke. Here were the variations which gave natural selection its power.

However, in Fisher's own words, the argument was 'rather complex', and the language of mathematics defeated at least one scientist

THE LYSENKO DOCTRINE

From the end of the Second World War to the Kruschev era, Mendel's genetics were rejected in Russia. Enthroned in their place were the ideas of I.V. Michurin, taken up and promoted by a famous protégé of Lenin, T.D. Lysenko. The disastrous harvests at the end of the 1920s had sent Soviet agriculture on a desperate search for ways to improve grain yields. The genetics then developing in the scientific world offered no immediate help.

Lysenko, a plant-breeder in Azerbaijan, achieved fame by showing how wheat seed which was normally sown in the autumn could in fact be planted in the spring. The crop could then be harvested later in the summer and would not be required to survive the rigours of the Russian winter. The trick was to allow the seeds to take up water, keep them just above freezing-point for a few weeks, and then allow them to dry out. As a result they matured more quickly when sown. Here was a positive benefit compared to the purely theoretical talk of genes, particles which no one had even isolated.

This tampering with the natural life-cycle of seeds would have had little relevance to genetics if Lysenko had not insisted that these changes, once induced, could be inherited. Western geneticists were confident that freezing and drying the seeds could not alter their genetic make-up. Whatever the benefits of treating the seeds as Lysenko did, it would have to be done for each generation; the ability to grow quickly would not be passed on. But Lysenko disagreed, and adopted a simple form of Lamarckian inheritance of acquired characteristics. And, fortunately for him, Soviet ideology was on his side.

Marxist-Leninist thought has always given pride of place to the influence of the environment. Change the conditions and the manner in which people work and you will change the people. Since this was fundamental to the way Marxists viewed the world it seemed appropriate that the same should be true in biology: change the conditions under which the wheat seeds were handled and the seeds themselves would change — for good. Lysenko also claimed that changes in the chromosome arrangements of wheat strains could occur 'by a leap'. This was wonderful news for leaders of a nation which in the Revolution of 1917 had suddenly leaped from feudalism to become a new socialist state. Nature and society were in harmony, truth in one was being reflected in the other.

assessing his work. The Royal Society refused the paper. When eventually the work was published in 1921 it was quickly recognized as seminal. Fisher had bridged the gulf separating the two sides: the observations of the statisticians were seen to be explainable by the geneticists, and vice versa.

Further work by J.B.S. Haldane in England and Sewall Wright in the United States confirmed and enlarged Fisher's ideas. During the 1930s the synthesis became closer. Theodosius Dobzhansky's *Genetics and the Origin of Species* (1937) is seen by many as the first real bridge between the experimental geneticists and the older style naturalists. Here gene recombination and mutation was understood as the source of the small variations needed to fuel the engine of evolution by natural selection. Darwin's insistence on the accumulation of tiny variations was reinstated.

The opposing camps met in the middle. The naturalists conceded that inheritance did indeed work with discrete particles, and abandoned their discipleship of Lamarck. The geneticists agreed that the effects of individual gene changes were small, and that sudden mutations were not the driving force of evolutionary change. As they looked up from their isolated 'freaks', and considered whole populations of animals, they returned to a belief in natural selection for the origin of species. Genetics became the study of the mechanism whereby variations occurred and were transmitted from one generation to another. From then on natural selection took over.

Apart from occasional notions of evolution being seeded from outer space or proceeding by 'hopeful monsters', the orthodox scientific view once more settled down to a modified Darwinism. In 1942 Julian Huxley, grandson of 'Darwin's bulldog', could write his major work and entitle it *Evolution: The Modern Synthesis*.

8
Understanding the Biology:
DARWINISM REFINED

The discovery of the structure of genetic material may be ranked in importance alongside the discoveries of Darwin or Mendel. Today we are entering an age when biotechnology can manipulate genetic structures with the same ease as the earlier industrial pioneers harnessed the power of steam. But we are only now enjoying the fruits of careful research conducted during the first half of our century. It was not so long ago that the very existence of genes was doubted!

With the unravelling of genetic structures has come a new understanding of heredity. And with this understanding came first confirmation and then doubt as to the validity of Darwin's natural selection. The picture of evolution is far more complex than was first believed. Thomas Huxley would be hard pressed to read a recent scientific paper and conclude, 'How extremely stupid not to have thought of that!'

Alongside development in science and technology our Western society has also made considerable movements along the moral and social line — some call this progress and others degeneration. The power and prestige of science has made many forget God and a new challenge to evolution has arisen from religious believers who see it as a destroyer of the soul. As one angry New Yorker wrote to a national magazine:

> 'Flushed with success in having proved that the physical earth was round, science has arrogated to itself the power to conclude

'Escherichia Coli' bacteria multiply at a quite impossible rate. There has to be some natural way by which such a population is controlled.

that the spiritual earth is flat, and the insidious hypothesis of Darwinian evolution is its chief argument in this campaign.'

Once more we are immersed in military invective and the ghost of Dayton past. But first we must look at genetics.

The nature of genetic material

Since the middle of the last century it has been known that cell nuclei contain nucleic

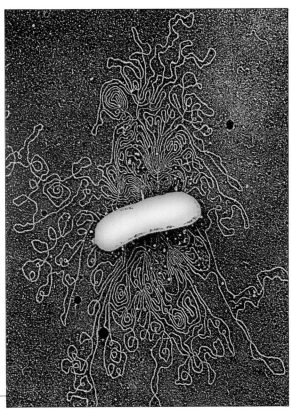

acid, a previously unknown substance rich in phosphorus. When he discovered the new acid the Swiss chemist, Friedrich Miescher (1844-95), was not looking for the mechanisms of inheritance. He had simply been advised by his uncle to study the chemistry of cells. And so, following his graduation, he went to work under the famous organic chemist Hoppe-Seyler in Tübingen. Further experiments revealed that chromatin — the material in the chromosomes — behaved in similar ways to nucleic acid, but Miescher did not see his new substance as linked to processes of inheritance. As Darwin had turned away from his central studies to a detailed examination of barnacles, so Miescher spent years examining the life history of salmon, the chemistry of sperm, nutrition in Swiss federal institutions, and variation in human blood chemistry as climbers ascended the Alps! Only much later in life did he return to his nucleic acid research. Then he died within a short time of tuberculosis.

By the turn of the century biochemists had established the presence of nucleic acid in plant and animal cells. There appeared to be two types: Ribonucleic acid (RNA for short) and Deoxyribonucleic acid (or DNA). There was an initial confusion as to whether or not one type of acid was for plants and the other for animals. But by the 1930s it was established that plant and animal cells contained both acids, the DNA form being located almost exclusively in the chromosomes whereas RNA existed in the cytoplasm (the protoplasmic content of the cell outside the nucleus). But still nobody knew what DNA was for. Its place in the chemistry of life was very unclear. If anything, its role was thought to be minimal. The apparently simple structure of DNA made it an inadequate chemical to contain all the necessary hereditary information, and the complex protein molecules were viewed as more favourable candidates.

How could chemicals in the chromosomes contain the detailed blueprints for life? Each parent must pass on to its offspring minute details on how to construct the specific proteins necessary for life. In an influential book *What is Life?* Erwin Schrödinger, by then a famous physicist, discussed the possibility of genes being coded information carriers. He reasoned that the only way in which genes could carry their hereditary information was by encoding the details in repeating strings of molecular structures. Just as morse code can express letters by sequences of dots and dashes, so genetic material might be made up of strings of molecules whose sequential order represented the information.

Of mice and DNA

The first step in understanding the chemical basis of inheritance came in 1928 when F. Griffith obtained some very strange results from an experiment he was performing on bacteria that cause pneumonia. These bacteria occur in two different 'phases', rough and smooth. These names relate to the appearance of the bacteria when they are grown in the laboratory on nutrient jelly. The 'rough' bacterium is not so virulent as the living 'smooth' bacterium and is not able to produce pneumonia.

At first all went according to expectations. If a mouse was injected with 'rough' bacteria

In his book 'What is Life', Erwin Schrodinger suggested how genes might carry their encoded information. Schrodinger was a physicist rather than a biochemist but his work pointed a way forward in genetics.

it did not become infected and all the bacteria died. If it received 'smooth' bacteria the mouse died of pneumonia and live bacteria could be recovered from the body. To confirm it was the bacteria which produced the pneumonia Griffith injected dead 'smooth' bacteria, and the fortunate mouse continued to scamper about its cage. It was definitely the living 'smooth' bacteria which produced the pneumonia.

Then Griffith had a shock. He injected both dead 'smooth' bacteria (which should be harmless) *and* living 'rough' bacteria (which he also knew to be harmless) and found the mouse had pneumonia. And in the corpse he found living 'smooth' bacteria! Either the dead smooth cells had 'risen from the dead' or the rough cells had somehow changed to become virulent, presumably because of some chemical transferred from the dead smooth cells. The second option seemed more likely, and so, in 1932, J.L. Alloway treated rough cells with a sterile filtrate from smooth bacteria and showed that this would also change the rough cells into smooth cells.

What was this 'transforming principle'? The answer came in 1944 when Oswald Avery and his co-workers in the laboratories at the Rockefeller Institute, New York, purified the substance and showed it to be deoxyribonucleic acid (DNA). If Avery was unsure of his result so were many of his contemporaries, and there was opposition to his conclusions until the early 1950s. It was still too difficult to believe that the simple structure of DNA could contain all the necessary genetic information. But Ernst Mayr records that 'the impact of Avery's findings was electrifying' and they generated an avalanche of nucleic acid research.

Chemical analysis of DNA showed it contained four chemical sub-units called bases. Initially it seemed that these bases regularly repeated themselves throughout the DNA molecule. However, as better analytical techniques were developed, it was shown that this succession of bases was not a monotonous sequence from one end of the DNA strand to the other. The bases could follow each other

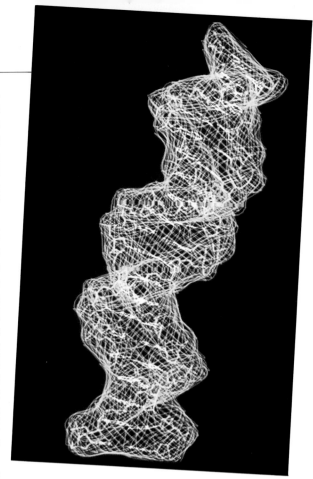

A computer-generated model of the DNA molecule shows the coupling between two strands within a single spiral.

in any order. It was immediately obvious that here was Schrödinger's code-script. The four bases were the dots and dashes (and pips and squeaks!) that made up the code of life. Alter the sequence of the bases and you altered the genetic coding.

The double helix

If the order of the bases expressed the life-code, what was the exact structure of the DNA molecule which allowed itself to be reproduced? Replication requires that an exact copy of the genetic code is copied so that it can be passed on to other cells and offspring.

In 1951 at the California Institute of Technology, Linus Pauling described the structure of a protein molecule. His success, in which guesswork and model-building played as much

MUTATIONS AND THE STRUCTURE OF DNA

DNA has four bases: normally an adenine base is linked to a thymine base by two chemical bonds and guanine is linked to cytosine by three chemical bonds. This means that it should not be possible for one base to be accidentally substituted for one of the other bases.

It should not be possible, but on occasion it does happen. All four DNA bases can temporarily exist in rare states which can join to the wrong base and form a wrongly matched base pair. For example, a rare form of adenine can link with cytosine instead of thymine. And once established in the DNA chain these incorrect links will be carefully reproduced each time the cell divides. The mutation will be passed on to all subsequent generations.

As well as substituting one base for another, 'deletions' and 'insertions' can involve the loss or gain of a single DNA base from the molecule. These changes garble the genetic code and a single amino acid may be altered in the protein being formed. Alternatively a completely different protein is manufactured.

Alteration in a single gene is not the only type of mutation. Changes can occur in chromosomes as well and these can be of six basic types:

☐ *Inversions* — a section of the chromosome remains in the same place but is turned end to end;

☐ *Translocations* — whole sections of chromosomes are exchanged between matching chromosomes;

☐ *Deletions* — a section of one chromosome is removed;

☐ *Duplication* — the chromosome receives two copies of a particular region of chromosome;

☐ *Fusions* — two whole, or almost whole, chromosomes join together end to end to form a single chromosome;

☐ *Unequal divisions* — one of the daughter cells loses one or more chromosomes during cell division and the other daughter gains extra ones.

If you think of the genetic information in a chromosome as the pages in a student's loose-leaf file then an inversion or translocation is like muddling up the pages, deletions are like ripping out one section and duplication is like copying the same notes out twice! Deletion of a section of a chromosome is usually fatal, but in the case of a duplication the additional genetic material may be an important source of evolutionary innovation. Lack of a chromosome is most often fatal, and even having an extra chromosome can produce severe disturbances. In humans, for example, Down's Syndrome is caused by an extra chromosome.

The majority of mutations, whether they are point mutations or chromosome mutations, have a harmful effect on the cells in which they occur. Very few mutations will be advantageous to an organism and promote its survival over the other members of the population.

Mutation is the only way new genes can be made but it is not the main way in which variability is maintained within a population. Every individual contains only a certain portion of the total genetic material in a population, and when two individuals mate the offspring contains a new mixture of that genetic material. An animal

a part as detailed analysis, inspired a young American to determine the structure of DNA. James Dewey Watson was then working in Copenhagen, but he moved to England where scientists were using the new techniques of X-ray crystallography on large organic molecules. Such techniques gave a picture of a molecule, but it took considerable skill to fathom out a three-dimensional organic structure from a crude two-dimensional representation. In England Watson met Francis Crick and sold to him the idea of tracking down the structure of DNA.

The story of their research, with its inspired guesses and wrong turnings, is a book in itself. They used the X-ray pictures of Maurice Wilkins and Rosalind Franklin at the University of London and they were guided by the requirement that the structure must be able to accommodate a coding sequence of bases. Without the ability to incorporate variable sequences of bases it could not act as genetic material.

In 1953 they published their now famous conclusion that DNA is a double helix — like two strands of rope intertwined — with the bases linking the strands together. DNA is like a twisted rope ladder, with each rung representing a pair of base molecules. Imagine such a ladder, with rungs painted in different colours — some green, some red, some blue, and some yellow. Each ladder could be made unique by altering

embryo contains equal numbers of chromosomes from both its parents. The mix of characteristics in the adult animal depends on which genes are dominant and which recessive.

But when the animal comes to produce its own sperm or eggs which chromosomes does it select? Will it use the genes from its mother or from its father? In fact it uses a mixture of both. It cannot use

Englishman Francis Crick and American James Dewey Watson were the first to establish the double-helix structure of DNA molecules. Great advances in biology and in medicine have followed their discovery.

both sets or the next generation will have double the number of chromosomes — one complete set of genes from each parent. The sperm or egg 'cells' are produced with half

the normal number of chromosomes and the selection is made by a process known as 'crossing over'. Matching chromosomes line up alongside each other and fuse at certain points along their length. When they are then pulled apart during cell division whole pieces of chromosome are exchanged, and the resulting genes of sperm or eggs are a mixture of the original matching pairs.

When you think of the many different hands that can be dealt from a pack of fifty-two playing cards, and then consider how many different combinations are possible with the thousands of genes which go to make up even the simplest living organisms, it is easy to see how so much variability can exist in a population. And that variability does not need a single mutation. But add the joker to the pack and the number of possible hands increases even more. Here, at the molecular level, is the origin of all the differences we see between individuals. This is why the baby 'has mother's nose and father's hair' and yet remains an individual in his or her own right.

the pattern of colours displayed by the rungs. On some a yellow rung would always follow one of blue, on others yellow might follow green. So in DNA the sequence of bases can vary. At one moment the bases adenine-thymine might follow guanine-cytosine and at another cytosine-guanine. In DNA the genetic-coding requirements are met. And the whole structure can be copied by unravelling itself into two separate strands like breaking apart the rungs of the rope ladder. The resulting severed links between the bases then pick up matching molecules and form two new strands. By this uncoiling, each spiral in the helix acts as a template for a new spiral. And because the bases can only fasten on to matching equivalents

the new manufactured spiral is identical. The mechanism can reproduce from cell to cell, from parent to offspring.

As Watson and Crick coyly state in their original paper:

> 'It has not escaped our notice that the specific pairing we have postulated (between bases) suggests a possible copying mechanism for the genetic material.'

There was still much work to be done. Watson and Crick had unveiled the coding machine, but what was the code? And assuming that the order of bases determined the proteins required for life, how did DNA actually produce these molecules?

This female fruit-fly has a genetic mutation which makes its eyes white.

The second question was answered first, and explained the purpose of DNA's cousin — RNA. In 1961 the French scientists François Jacob and Jacques Monod proposed that RNA acted as a 'messenger', taking a copy of the DNA instructions to particular sites in the cell where proteins were produced. The RNA moves through these sites (called 'ribosomes') much as a tape passes through the playing head on a tape recorder. The sequence is read off and translated into proteins.

A virtually unknown biochemist, Marshall Nirenberg, began the process of breaking the genetic code itself. He mixed messenger RNA with the protein-forming chemicals. However, the RNA he used was synthetically produced and did not contain the usual combination of four bases. It contained only one. The dramatic result was the production of proteins composed of chains of only one type of amino acid (amino acids are the building blocks for the more complex proteins). A flurry of further research across the world used the same techniques to identify the bases coding for all the twenty amino acids that make up our proteins. If you fed in RNA of a certain structure, a particular amino-acid chain would be produced. From these 1960 experiments have flowed our present knowledge of genetic materials and the ability to adapt the code for our own purposes.

Whatever uses we make of such knowledge there is no denying the awe-inspiring nature of the mechanism we have uncovered. No one can pretend that molecular biochemistry is simple, but it is certainly wonderfully elegant. Had Archdeacon Paley been alive he would have incorporated it in his book alongside his illustration of the human eye.

The doubts begin

These advances in genetics served to strengthen the neo-Darwinian synthesis. The scientific doubts about evolution by natural selection were by now all but silenced. At the centenary celebration of the publication of *Origin of Species* in 1959, eminent men of science said that if a layperson sought to criticize Darwin's conclusions he or she must be guilty of 'ignorance or effrontery'. Anyone who failed to honour Darwin inevitably attracted 'the psychiatric eye' to himself. These were strong endorsements for neo-Darwinism and a powerful warning against rocking the boat!

Thomas Huxley had warned Darwin that new truths often begin as heresies and end up as superstitions. By 1959 this was undeniably true of Darwin's theory of evolution. It was uncritically accepted as true by the bulk of the scientific community and taught as such. In the general enthusiasm for the New Synthesis there were only a few individuals swimming against the tide of orthodox opinion. And these could always be dismissed as mere eccentrics.

But recently more and more voices have been raised questioning Darwin, and though not yet a roar they certainly cannot be ignored. The academic credibility of these dissenters is so good that it is rash to dismiss them as ignorant upstarts or as people in need of psychiatric help. In all fairness to the orthodox view the questions now being asked are largely the result of our increased knowledge and understanding of the mechanisms at work in biological systems. This is knowledge that was simply not available in 1959.

The adherents of the creationist view try to make capital from what they see as the disarray in the scientific camp, using it as support for discarding evolution completely. Nothing could be further from the truth. There is in fact little disarray in the scientific camp at all. What has been happening is that the findings from several different disciplines within biology have shown that natural selection may not be the only mechanism involved in evolution. But of the fact that evolution has occurred there is no doubt, no disagreement.

Workers in fields as diverse as palaeontology, embryology, ecology and molecular biology now see evidence to suggest that the situation is more complex than neo-Darwinian theory allows. If they seek to do anything it is to make additions to Darwin's theory rather than to abandon it completely.

So just what are the scientific objections to neo-Darwinism? They may be listed under six headings, and we need to consider each in turn.

The fossil record

Over the period from the publication of *Origin of Species* to the present day there has been a tendency to interpret Darwin's idea of evolution as a gradual process. Species alter slowly by a series of small adaptive changes in response to their environment. One of the central beliefs of the Darwinian idea of evolution is this 'gradualism', although 'gradual' and 'rapid' are only relative terms as we shall see in a moment.

An implication of gradualism is that the fossil record should show slowly changing forms. All the intermediate stages of evolution should be buried as fossils and we should be able to unearth them. Now Darwin himself was at pains to explain that the fossil record was incomplete. The conditions under which a fossil can form are such that it is by no means a common event. Combine this with the fact that we only nibble at

the edges of the rock strata which have fossils in them and it is easy to accept that our knowledge of the fossil record is far from complete.

But most palaeontologists today would agree that the fossil evidence, rather than showing a nice gradation from one form to the next, shows fossils persisting throughout long periods of time virtually unchanged. At the end of such a series the change to a new form is 'instantaneous' — which in the geological time-scale of the fossil record can mean a period of up to 50,000 years. To give but one example, studies of marine bivalves, small clam-like creatures, from the Jurassic Period of between 135 and 180 million years ago, show that apart from increasing in size they hardly changed at all for periods of 10 to 12 million years. After this they were quickly replaced with a markedly different species.

Evidence like this led Stephen Jay Gould, a professor of geology at Harvard University, and Niles Eldredge, at the American Museum of Natural History, to propose the theory of 'punctuated equilibria'. As they put it, 'the

Stephen Jay Gould, professor of geology at Harvard University, has suggested reasons why the evidence shows evolution developing in sudden spurts rather than as a steady process. Gould was called for the defence in the Arkansas trial about evolutionary teaching in 1981. Photograph by Jill Krementz.

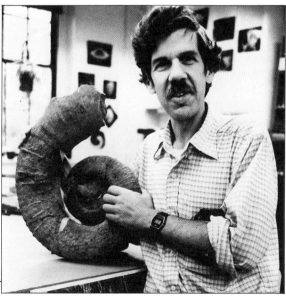

history of evolution is not one of stately un-folding, but a story of homeostatic equilibria, disturbed only "rarely" by rapid and episodic events of speciation'. In simpler terms, evolution progresses in spurts. It is like a marathon runner who, rather than running slowly all the way, covers the distance by a series of short sprints followed by long rests. To Gould and Eldredge the long periods of geological time where little change can be seen are not the product of an incomplete fossil record but are in fact an important feature of evolution. The reason we do not see the evolutionary changes recorded in the fossils is that for most of the time no change was taking place. On our marathon analogy, taking a photograph of the runner every ten minutes would produce only a few pictures in which he or she is moving. We would have to be lucky to catch the runner in one of the moments of sprinting. Where Darwin said that the fossil record was incomplete through lack of fossilization (the photographer never took a picture), Gould and Eldredge said fossils of intermediate forms are rare because of the speed of change (the photographer is unlikely to get a picture with the runner sprinting).

Such an idea is not really in conflict with Darwin's theory or with the new synthesis. The long periods without change in the fossil forms could be the result of stabilizing selection. And a period of 50,000 years, an instant of geological time, could cover well over 10,000 generations of most plant or animal populations on a biologi-cal time-scale. When Darwin was arguing for an evolutionary origin for the diversity of living forms on the earth he used the word gradual to counter the idea of a truly instantaneous creation of the kind envisaged by the creation-ists. The 'jerky' development as envisaged in punctuated equilibria is completely acceptable in the context of 'gradual' evolution as used by Darwin.

Of much more fundamental importance is an idea, which accompanies the punctuated equilibria theory, that the sudden changes from one form to another coincide with major cata-strophes. At each catastrophe large numbers of

species are wiped out. The fossil record shows evidence of mass extinctions. And it even enables us to place the six major ones as having occurred at the start and end of the Cambrian Period (600 million and 500 million years ago respectively), and the ends of the Devonian (345 million years ago), the Permian (225 million years ago), the Triassic (180 million years ago) and the Cretaceous (63 million years ago). It has been calculated that up to ninety-six per cent of all life forms were destroyed in these six cataclysms. That is, only four per cent survived.

Evolution by natural selection relies on changes in the environment which are small enough for the individual organisms to with-stand, at least until they can produce the next generation. If the change in the environment is too great and too rapid for the organism to survive and reproduce, there will be no subsequent generations. The question as to how well adapted to that environment the individuals of the next generation are just does not arise. This is the situation with the six global catastrophes mentioned. No well-adapted next generation was ever formed.

In the catastrophes we are describing the changes were outside the adaptive ability of most of the organisms then inhabiting the earth. Survival was more a matter of chance than of adaptive superiority, more a matter or luck than of design. But once the catastrophe was finished the world in which these fortunate survivors found themselves was less competitive. There were fewer organisms to share the plenti-ful resources. There were many empty niches into which the plants and animals could expand and conditions were right for adaptive radiation to occur.

So, the main trends in the evolution of living forms have been determined by a series of uncoupled events — by mass extinctions rather than by a slow progression of many small steps. Here we have a stark contrast with the Darwininan or neo-Darwinian theory of evolution. This whole concept has been called 'neo-catastrophism' as it shows marked similarities to the views of the geologists of the early nineteenth century. They believed that a series of cataclysms had produced the fossil formations which we now see.

Parallel evolution

The fossil record has provided a picture of the development of life on earth which is at odds with the neo-Darwinian concept of evolution. And there are more difficulties to come. A further phenomenon mentioned by those scientists who have doubts about Darwinism is that of 'parallel evolution'. Here two groups of animals, widely separated from one another on the globe, show remarkably similar evolutionary histories.

Take, for example, the striking similarity between the marsupial animals of Australia and the placental animals of the rest of the world. Both groups are mammals and had a common ancestor — a small shrew-like creature which lived on earth at the same time as the dinosaurs, some 150 million years ago. As the earth's crust shifted and the land masses separated from each other, a small group of these mammals became isolated on the Australian sub-continent. This population diverged from the usual mammalian way of nurturing the developing embryo — by a placenta linking the mother's blood supply to that of her offspring. Instead the progeny were born very early and raised in a pouch on the mother's belly. Food was supplied from a teat inside the pouch. What is so curious about the marsupials is that, although they have evolved for millions of years in isolation from the remainder of the world, they have produced look-alike equivalents for many of the placental mammals.

The Tasmanian wolf, or *Thylacine*, now sadly on the verge of extinction, is remarkably similar to the Canadian timberwolf. Is it simply, as the neo-Darwinists say, that the two species live in similar environments and occupy similar ecological niches so that they have been subjected to similar selective pressures? But then think of all the potential variations possible in these two animals and the differences which

must have occurred between their environments, both living and non-living, during the different stages in their evolution. It seems highly unlikely that the similarity is solely due to natural selection of random mutations! It seems far more likely that the similarity exists because there is some basic pattern for 'wolves' which determines the overall form they take. The genes of the two isolated species can only produce variation within the limits of a basic pattern.

The variation which is achieved involves a complex interaction between organism and environment. And it works both ways. We tend to forget that the activity of living organisms also changes the environment. Not only are plants and animals adapted to their environment but their environment, even the inorganic part of it, has been adapted by them for the support of life. It is as if the environment provides a template within which a restricted number of evolutionary routes can be followed, but each step along that route causes a change in the template.

James Lovelock, an independent scientist and visiting professor at the University of Reading in England, has incorporated the idea of this reciprocal relationship into what he has called 'the Gaia hypothesis'. Named after the Greek earth-goddess, this hypothesis sees the evolution of living organisms to be so closely coupled with the evolution of their physical and chemical environment that they are really a single, indivisible evolutionary unit.

Hopeful monsters

We have already mentioned the work of Richard Goldschmidt. He was a refugee from Hitler's Germany during the 1930s and 40s who worked at the University of California in Berkeley. He discovered a number of evolutionary features which he could not accept as the result of gradual and continuous change. Though Goldschmidt accepted the neo-Darwinian idea of gradual change *within* a species, he could not believe it was responsible for changes *between* species. He proposed that these changes happened suddenly by a massive mutation to form 'hopeful monsters'. Of course most of these died as they were unsuited to their environment. But now and again one came through to become the founder of a new species.

A major objection to the idea of a hopeful monster is that a single individual will have little effect on a whole population. If it is a sexually reproducing organism and the mutation is such that it can no longer mate with the parent population, then the new species stops there! Even if it could still mate with members of the parent population its unique contribution of genes would be swamped by the others. You need a number of such monsters produced at a single time before there is any effect on the population. Is that possible? Well, the answer, surprisingly, is yes!

The development of an organism from a fertilized egg, through an embryo, to an adult form is a carefully controlled process. It leads to a predictable result. Conrad Waddington of Edinburgh University uses the metaphor of hills and valleys forming an 'epigenetic landscape' through which the developing embryo 'rolls'. Epigenesis means the formation of an entirely new structure during development. Normally the embryo takes the line of least resistance and rolls along the bottoms of the valleys to the predictable outcome of the adult form. A duck egg becomes a duckling, which in turn become a duck.

But if the landscape receives a 'blow' while this is happening then the embryo may be jolted out of the usual valley and over a hill into a different valley. The outcome of this could well be a totally new adult form. Now what do we mean by giving the embryo a blow? If you give embryo fruit-flies a sharp heat shock or treat them with chemicals, they emerge from the egg in a different form. Different shocks produce different outcomes, but at least here we have a number of hopeful monsters produced simultaneously, and such changes can become established in the whole population. If some theories about the extinction of the dinosaurs harked back to nineteenth-century catastro-

phism, then Goldschmidt's theories contain shades of *Vestiges*!

During its embryo stage an organism is at its most susceptible to change. Its cells have not yet become fixed in the roles they will eventually play, and relatively small changes here will produce much greater changes in the final adult. Goldschmidt was one of the first scientists to suggest the role played in the development of an embryo by genes controlling the rate of developmental processes. For instance, he noted that quite small differences in the timing of pigmentation in the embryo resulted in large differences in the colour patterns of full grown caterpillars.

D'Arcy Thompson has shown that a geometric relationship exists between the shapes of different animals. If you place a grid of lines, like a piece of graph paper, over the image of one animal, then by a simple distortion of the grid pattern it is possible to create the form of a different animal. What rules underlie this transformation we do not yet know, but the fact that it is possible to model the process with

Everyone knows that ostriches have calluses on their knees. These calluses thicken as the ostrich kneels on them, but this thickening begins before the ostrich is born.

fairly simple geometric methods indicates that it is not random.

One of the obvious candidates for a biological mechanism controlling such processes of embryonic development are the master-control, or homoetic, genes. A single mutation in one of these genes produces massive changes in the body form of the organism. To give an example from the much studied fruit-fly *Drosophila*, mutations of one homoetic gene can cause part of the fly's head to develop as if it were part of its thorax. Legs grow out where there should be antennae! In such a bizarre case the emphasis is on the 'monster' rather than the 'hopeful', but it illustrates quite clearly that sudden and dramatic body changes are possible as the result of single genetic mutations and that Goldschmidt's 'hopeful monsters' are not so far-fetched after all.

A look at the ostrich's knees

This all seems fine, as far as it goes. But what about the problem of the inheritance of these characteristics? Even if a viable monster is formed, how are the genes transmitted to the next generation? The germ cells separate off from the body cells at a very early stage in embryonic development. Any characteristics induced in the body cells by the environment after this point cannot be passed on to the next generation. This is known as 'Weismann's Barrier' and is one of the central dogmas of modern biology. Environment cannot influence the characteristics of the germ cells. The cells might become damaged themselves by X-ray radiation, but they cannot pick up changes imposed on the adult. Or can they?

In June 1970 the scientific journal *Nature* contained two papers, one by David Baltimore and one by Howard M. Temin and Satoshi Mizutani. These showed that the central dogma of molecular biology, that information only passes from germ cells to adult cells and never the other way, was not universally true. Both research groups, working independently, discovered an enzyme in certain tumour viruses which could cause the process of 'reverse transcription'. In normal transcription, RNA (which is the cell's humble messenger chemical) copies the coded message of the DNA and takes it to the site where new proteins are produced. The scientists had discovered how to trigger the reverse process where the coded message in the RNA of the virus was transcribed into a DNA strand. This in turn then became spliced into the DNA of the hosts' cells. The class of viruses which are capable of doing this are now called 'retroviruses' and probably the most widely known example is that which causes AIDS.

Since 1970 the revolution of genetic engineering and the advances in recombinant DNA technology have routinely used this process under laboratory conditions. But research now shows that there are molecular mechanisms in the cells of plants and animals whereby such reverse transcriptions can occur quite frequently under natural conditions. So can changes in body cells affect the germ cells? And can these new germ cells determine future generations?

In the 1950s British biologist Sir Peter Medawar studied the immune system in mice, that is, the mechanism which protects them against the invasion of foreign material such as bacteria. The cells of the immune system recognize the tissues of their own body and can distinguish a foreign tissue or protein from 'self'. If something goes wrong with this recognition of 'self' then auto-immune diseases such as rheumatoid arthritis can occur. Here the immune system cells attack and try to destroy the tissues of their own body. Medawar found that in a mouse the immune system is still very flexible for a short period after the birth and a massive injection of foreign material during this stage causes no ill effects. Subsequently the foreign material is regarded by the immune system as part of 'self' and the adult mouse is tolerant to that foreign material.

Almost three decades later Ted Steele, Reg Groczyski and Jeffrey Pollard, of the Ontario Cancer Institute in Toronto, used this fact to test Weismann's Barrier. They injected millions of foreign cells into the bodies of newborn mice. As Medawar had found, the mice became tolerant to the invading cells. In some way the genes of the body cells of the mouse had become altered so that they did not reject the foreign cells, and this new tolerance remained for the lifetime of the mice. But the researchers then went on to show that this tolerance could be inherited! Some of the children and grandchildren of the experimental mice also tolerated the foreign cells. And the numbers of tolerant mice in each generation were in the ratios predicted by Mendelian genetics.

Studies of this kind have increased speculation among some scientists that Lamarck was not completely wrong after all. The inheritance of acquired characteristics could also explain why the ostrich has callouses on its knees where it kneels on the ground. Certainly the callouses are thickened because the ostrich kneels on them. But the intriguing thing is that the thickening begins before the ostrich is born. Perhaps the explanation lies in the neo-Lamarckian application 'of the molecular mechanisms of reverse transcription.

Drifting genes

The emphasis in neo-Darwinism is that all changes occur under selective pressure. No change occurs which does not offer some evolutionary advantage or is linked to one that does. But research by the Japanese scientist Motoo Kimura has now questioned this.

Motoo Kimura has shown that the rate of point mutation is much the same in all animals but that this is not reflected in their rates of evolution. Different groups of animals evolve at markedly different rates. Kimura suggests that a large number of mutations are neutral in their effect — they confer neither advantage nor disadvantage to the possessor. Such neutral mutants may become fixed in the population by a random process of genetic drift.

Imagine that you have a large jar of jelly babies. From the jar you take out several separate handfuls of sweets. When you count

the number of different coloured jelly babies in each handful you will find that they are not in the same proportions. In some handfuls there may be one or more colours missing, in other handfuls the sweets may be almost all one colour. Compared with the overall proportions of the different colours in the full jar none of the samples corresponds exactly. They all show some 'sampling error'.

Each generation of an animal population is a sample of the total genetic information in the gene pool. And, like our random handful of sweets, the frequency of the various genetic forms varies from generation to generation. If by chance, through sampling error, the frequency of one form is increased then this improves the likelihood of that particular genetic form being passed on to the next generation. But, unlike our jelly babies, the effect may be cumulative through the generations until only the one genetic form remains. It is now fixed.

The reverse is also possible. A series of decreases in the frequency of a certain gene form could remove it completely from the gene pool. As the sampling error is random it could just as easily increase the frequency in one generation and decrease the frequency in the next generation, perpetuating changes in each succeeding generation. These changes will continue until the gene form is either fixed or lost from the gene pool. Now it is important to note that changes in genetic structure may become fixed, but they originally arose quite by chance. There was never any pressure from the environment to prefer one genetic form to another.

Another recent discovery is of 'pseudo' or 'dead' genes which have lost their function. Research has shown that the rate of mutation within a pseudogene is much greater than the rate of mutation in the similar, but functional, gene. For example, the normal gene for globin, a protein associated with the oxygen carrying red pigment in blood, changes DNA bases at only one tenth the rate of the globin pseudogene.

At first sight such a thing is difficult to explain by natural selection. But if we remember that all the changes in the dead gene are neutral, whereas in the functional gene a number will be harmful, then it makes sense.

The large reserve of non-functional DNA makes it possible for mutations to occur which remain substantially hidden as far as the characteristics of the grown animal are concerned. Scientists are only just beginning to understand the mechanism by which genes are switched on and off, but it is possible that the simultaneous activation of several pseudogenes which contain previously neutral mutations could allow the sudden changes we find in the fossil record.

Sociobiology

Within the populations of higher animals we can recognize a definite social organization which serves to regulate the functioning of the group as a whole. Your own pet cat reacts differently to neighbouring cats as it seeks to establish its own territory. Such social behaviour has led to studies in the field of sociobiology. Here selective advantage is related to certain behavioural traits in individuals.

Some of these behavioural patterns are undoubtedly learned. For example, the blue tit or titmouse learns to open the metal foil tops on milk bottles to get at the cream. Some studies of Macac monkeys on the islands off Japan showed how one monkey, called Imo, started the habit of washing potatoes left out as food. Soon this habit had been 'taught' to a number of other monkeys in the troop. They too began to wash their potatoes in the sea. Transmission of cultural traits such as these throughout a population resembles the transmission of genetic traits through the population. Should a particular behavioural trait give a selective advantage to the individual then natural selection will favour that individual. A blue tit filled with cream may live longer than his hungry fellows. And if our intelligent blue tit passes on her skill to her offspring then that particular genetic group will flourish. Evolution will proceed more by cultural adaptation than by natural selection.

A lioness gently carries her cub. Lionesses will put themselves in danger to protect their offspring, and this is an example of apparent altruism within the animal world. But can this be explained genetically?

But consider the reverse. Are what we fondly call cultural traits really learned, or might they be chemically determined? Are we generous because we have learnt to be, or because we are programmed? Some sociobiologists have suggested that altruistic behaviour, sacrificing yourself for the good of others, is simply a biological expression of your genes. The bird which cries out when it sees a predator, so warning other birds yet drawing attention to itself, is not being brave or self-sacrificing. It is merely responding to pre-programmed genetic behaviour.

To sociobiologists the most important aspect of life is the transmission of the gene into the next generation. How that is achieved is of lesser importance: the body of the animal or plant is merely a vehicle in which the 'selfish gene' is carried round. It is fair to point out in relation to this last aspect of sociobiological thought that no gene for behaviour has so far been identified. But the cultural transmission of behaviour can and does have profound effects on the individuals in a population.

'Darwin's Death in South Kensington'

By now your mind is probably in a whirl. Just when the Darwinian theory of evolution seemed to be so soundly based you find there is evidence which, at the very least, topples it from a position of pre-eminence. Are the scientists wrong? Which should we believe? What can we believe?

It is well to remember that these scientists are in no doubt at all that evolution has occurred. But the mechanism by which evolution has occurred is in dispute. None of these scientists would go so far as to say that Darwin got it completely wrong. The worst they would say was that he did not get it all right. Darwin's original ideas have provided the framework within which much biological research has been conducted, and without the stimulus of his theory our knowledge of the plants and animals of the world would be that much poorer.

But the question still remains: what can we do in this state of confusion over the mechanism of evolution? The British Museum of Natural History had to face the same problem and it chose to do so by its view of 'cladistics'.

The Greek word *klados* means a twig or small branch and from it is derived the word cladistics. This is simply a way of classifying the living world according to observable features. By identifying the number of common features between two animals it is possible to see how closely related they are to one another. The results are presented in a diagram like the branching of twigs on a tree with each species as the tip of a twig. But unlike the older evolutionary trees no attempt is made to put a species at the point where two twigs divide. These

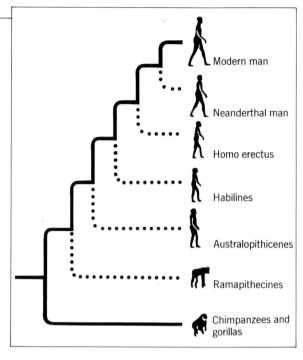

A cladogram makes no attempt to build ancestral relationships between species. It simply records the number of characteristics a creature shares with similar species.

diagrams do not trace the history of descent. They merely show the similarities between one species and the next.

It is a method of classification which intuitively feels right for we can easily accept that a dog is a closer relative of a cat than of a bird or a fish. With this approach there is a return to basics, an attempt to understand the pattern produced by evolution rather than speculating about the flow of genes from one population to another through time. The cladist's tree does not show which animal descended from which; it only shows which animals are most alike. And that is completely different. It is a twentieth century equivalent of the classifications of Aristotle or Linnaeus. Though it is accepted that the present diversity of living forms arose by evolution, the mechanism by which a particular species evolved is not relevant in cladistics. Indeed it is assumed that none of the species considered are ancestral to any other species. Colin Paterson, one of the staff of the museum, explained in a radio interview:

'As it turns out, all one can learn about the history of life is learned from systematics, from the groupings one finds in nature. The rest is storytelling of one sort or another. We have access to the tips of the tree; the tree itself is theory, and people who pretend to know about the tree and to describe what went on — how the branches came off and the twigs came off — are, I think, telling stories.'

And so the British Museum of Natural History side-stepped the scientists' doubts about evolutionary theory. But in so doing they upset a number of people. Under the headline 'Darwin's Death in South Kensington', a critical editorial in the scientific journal *Nature* described the museum's newly arranged exhibitions as being 'shot through with heresy'. Opened a year before Darwin's death, the British Museum of Natural History had been a citadel of Darwinism. In 1981 this was no longer true. The exhibits, replete with cladograms, were not suggesting any particular form of evolution at all!

A MARXIST PLOT?

The British Museum (Natural History) occupies a magnificent neo-gothic building in central London. Its towering church-like windows and architecture have moved many to characterize this site as a temple to Darwinism, with its exhibits forming the holy relics of a sacred scientific theory.

But in 1981 it lost its faith. Or so it seemed to devotees on the outside, who saw the new exhibitions being displayed as selling out to anti-Darwinian pressure. Criticism of the museum's policy came in two parts:

□ The museum decided not to display some of its priceless collection of preserved organisms, but instead to mount exhibitions on key aspects of the natural world. The first two exhibitions were on dinosaurs and on mankind's place in evolution. Their aim was to attract young people to the museum by providing interesting displays to which visitors could react by pressing buttons to check their answers against questions and puzzles. But critics wanted to know whether this was a wise undertaking for a museum. Education demanded adopting certain viewpoints — whether this fossil was an ancestor of that, or whether evolution was true. And when the exhibition guidebooks used phrases like 'If the theory of evolution is true . . . ', and employed cladograms (which ignore all questions of evolutionary descent) to express the links between hominid fossils usually thought to be ancestral to mankind, some scientists thought they had gone too far. The exhibition and its guidebook was a Doubting Thomas newly turned teacher. 'But is the theory of evolution still an open question among serious biologists?' asked *Nature* magazine. 'And, if not, what purpose except general confusion can be served by these weasel words?'

□ The museum's use of cladograms also betrayed a further fault, said Professor Halstead of Reading University. He emphasized a link between cladistics (which represents inter-species relationships by a branching diagram) and Stephen Gould's theory of punctuated equilibria. The diagrams show species on different 'branches' and, Halstead believes, they support Gould's idea that changes from one species to another occur very quickly. Stephen Gould has since denied this link, but the supposed connection was important for Halstead as it made the museum's exhibits propaganda for Marxism! And this worried Halstead greatly.

In his paper outlining punctuated equilibria Gould had pointed out parallels between the fossil record, which shows sudden jumps forward rather than steady progress, and the Marxist view of history which sees societies progressing in sudden stages via revolutions. The classical Darwinian position is that change is gradual, and Gould queried whether this viewpoint was, in part, a reflection of the political ideas of Victorian society. As Gould says in another essay, 'gradualism is a culturally conditioned prejudice, not a fact of nature.' Our beliefs away from the laboratory influence our scientific theories, and in fairness Gould mentioned that his theory of evolution by jumps could be seen as reflecting Marxism. With this in mind, Halstead concluded his letter,

'What is going on in the Natural History Museum needs to be seen in this overall context. If the cladistic approach becomes established as the received wisdom, then a fundamentally Marxist view of the history of life will have been incorporated into a key element of the educational system of this country.'

In short, the exhibitions were a sell-out to creationists, and the cladograms were Marxist propaganda. There was heresy among the faithful.

Again, Paterson explained:

'Just as pre-Darwinian biology was carried out by people whose faith was in the Creator and His plan, post-Darwinian biology is being carried out by people whose faith is in, almost, the deity of Darwin. They've seen their task as to elaborate his theory and to fill in the gaps in it, to fill in the trunk and the twigs of the tree. But it seems to me that the theoretical framework has very little impact on the actual progress of the work in biological research. In a way some aspects of Darwinism and of neo-Darwinism seem to me to have held back the progress of science . . .

'There is an extraordinary ferment going on within evolutionary biology at the moment. Where it will lead I wouldn't pretend to guess. I think that the general theory — that evolution has taken place — will remain, but that more people may come to realize that it is not essential to doing biological research to believe in it.'

Even though Paterson does not represent the majority, the adoption of cladistics by the museum has made many scientists re-examine their dogmatic acceptance of Darwinism. This re-evaluation is no bad thing. The strength of science has always been its willingness to subject even its most cherished beliefs to discussion and debate — and, if need be, to modify such belief in the light of new evidence. So it is with Darwin's theory of evolution. The final outcome of this debate is as yet undecided, but for the present the words of the Roman writer Horace seem appropriate, 'Scholars dispute, and the case is still before the courts'.

The rise of scientific creationism

Judgment on the validity of Darwin's evolutionary mechanism is tossed to and fro in the academic journals of our day. But evolution itself has also come to be questioned. A number of engineers and scientists have recently re-opened the issues of the Scopes trial — was the world created according to a literal reading of Genesis? For the beginning of the story we must go back to 1957.

In 1957, the Russians launched their first space vehicle, Sputnik. It brought not only the beginning of the space race but a reappraisal by the Americans of their science education. They were embarrassed by the success of the Russians. How had they achieved this lead in technology? Certainly if America was to gain lost ground then the next generation, it was argued, must be better taught and trained in scientific disciplines. Curricula were reviewed, and new textbooks produced. The aim was to bring science teaching up to date. The advances in biology during the 1940s and 50s had not filtered down to the classroom. For many schools this brought their first exposure to evolution and the recent theories on the development of mankind.

The National Science Foundation funded several programmes designed to modernize the teaching of science in American schools. The Biological Sciences Curriculum Study worked with scientists and teachers to develop a series of biological texts. Although these covered a wide range of biological subjects, they incorporated the theory of evolution as a major theme.

The project on *Man — a Course of Study* immediately ran into trouble. Started in 1963, it was based on the fundamental belief that all cultures were equal. Through gradual development humanity had come to adopt a whole series of personal, social and political values. For the project there were no fixed and absolute beliefs, nor was one culture superior to another. This the project writers saw as the social correlative of biological evolution. If animals are always changing, then cultural values must also be allowed to change alongside. If animals are evolving in response to the environment, then our human traditions will also change in response to alterations in the social 'climate'. We have already seen how the change from a rural to an industrial economy in the south of the United States fuelled the evolution debates. And, of course, if everything is diverging and

A model of the 1957 Soviet Sputnik, which 'won the race' into space. This setback caused the Americans to rethink the methods used in science education in schools, and teaching about evolution was one area to be drastically revised.

changing there is no moment at which you can say, '*this* is the age of true morals and judgment'. Everything is relative.

All this conflicted sharply with traditional beliefs. A religious understanding, centred on literal belief in Genesis, sees humanity and the animals created in their set forms at the beginning of time. The laws of behaviour (the Ten Commandments) were revealed to Moses, and written down on stone. Everything, our relationships and our rules for living, was fixed way back in history.

But now the biologists were describing a different scenario for our origins. And if animals were not created in their present form at a single, instantaneous moment, then (so the argument went) fixed, 'once-given' moral positions were not valid either. If one could develop over time then so could the other. Change was the name of the game. Indeed, since in the animal kingdom biological change and adaptation was often advantageous, then cultural change was the way forward for the human race. We needed to adapt our beliefs and values as conditions changed.

The argument, then, for believing no one

culture to be 'best' was rooted in our biological history.

To deduce social beliefs from biological observations has always been a dangerous pastime. We saw in the last chapter how nineteenth-century businessmen justified their own wealth and their neighbour's poverty by appealing to evolution. Hitler used the concept 'survival of the fittest' as the rationale behind his death camps.

But all this was a long way from the proposed school textbooks. The project advocated very humane and laudable ideas. It breathed tolerance for those of differing cultures. Many were happy to see such principles taught in their schools.

Yet they were grounded on the same faulty logic. For if our values are only based on biology then they can change as biological theories come and go. An influential group of scientists has only to 'discover' that mankind reached the pinnacle of the animal kingdom by savagery and deceit to introduce a new moral 'norm'. Since at present the emphasis within anthropology is on the benefits of co-operation, all is well. But we must be wary of the science-to-morals road. Mainstream Christianity, while accepting evolution in the biological sense, has always looked to God as the well-spring for treating others as equals. Undoubtedly cultures ebb and flow, but the Christian still finds the bedrock of morals in the eternal God.

This link between a description of human origins (evolution) and a prescription for human behaviour (ethics) lies behind much of the rhetoric of the 'Back to the Bible' campaign. Men such as Bryan believed that moral standards were falling, and that society was becoming increasingly secular. Evolution, because it appeared to provide an alternative perspective from which to view the world, was to be feared. It undermined stable values; it was shifting sand. It destroyed faith in God and the Bible. As the *New York Times* of 1922 quoted Bryan:

'They (believers in evolution) weaken faith in God, discourage prayer, raise doubt as to a future life, reduce Christ to the stature of a man, and make the Bible a "scrap of paper". As religion is the only basis for morals, it is time for Christians to protect religion from its most insidious enemy.'

Even though Bryan was right, from a religious viewpoint, in assessing the basis for morals, he was wrong in identifying the enemy as evolution. Just as science makes a poor foundation for morality, so concern for the moral standards of society is no reason for attacking biological evolution. Consciously and unconsciously we link our understanding of our

Adolf Hitler applied the evolutionary idea of 'the survival of the fittest' in a distorted way to justify death camps.

origins to our hopes for the future. As we shall see in the next chapter, there are some who have tried to set up evolution as an alternative to belief in God. But they are as much off target as Bryan.

We cannot be naive. Any theory of origins tends to generate an outlook that pervades our whole lives. And evolution suggests that change is beneficial, that nothing is fixed, and that God is redundant. In this sense evolution has influenced the way we view ourselves and value our neighbour. Christians may want to challenge some of these philosophical spin-offs. The church may want to reassert that God is the ultimate source and definition of what is right and good. But this is religion and philosophy, not science. In itself science merely plots the lines of cause and effect.

It is because the science of evolution has become entangled with deeper human questions of morality and belief that it has raised such a storm. We must try to separate it, although the task may be impossible. But we must try. Just as it is unfair to promote evolution (as a science) because we favour liberalism in morality, so it is unnecessary to attack evolution (as a science) because we believe in more fixed values. And yet it is only logically unfair and logically unnecessary. For while theories of origins touch our spirit so deeply the entanglement will remain. The best we can hope for is a clearer vision of why each one of us reacts to this seminal scientific theory in the way we do.

During the 1960s and 70s, a number of groups were formed in the United States in order to combat the teaching of evolution. They did not see the scientific theory as neutral, but as the cause of America's ills.

In 1963 the Creation Research Society was founded by ten disaffected members of the American Scientific Affiliation. This latter organization was comprised of scientists committed both to the Bible and to science, and it increasingly accepted evolution as God's method of creation. For some this 'theistic evolution' was a sell-out to secularism. The breakaway Creation Research Society

IS CREATIONISM ANTI-SCIENTIFIC?

Some might find it surprising that there are so many engineers and scientists who support creationism. Is not the supernaturalist approach of creationism anti-scientific?

No. There is a close harmony between a literalist view of Scripture and a straightforward approach to science. In both realms the creationist is happy to accept the 'givenness' of the subject.

In science he or she accepts clearly defined laws and a sovereign God. In Scripture he or she accepts the clear words of revelation. Both science and Scripture are realms of the given. They come from God and are accepted as reality. Where we can see clear laws in nature (such as gravity) we record them. Where we can see the clear meaning of the Bible, we apply it. And when it comes to origins then the Bible gives details of those first creation days, days which science can never go back in time to observe first hand. The literal meaning of Genesis 1 should be trusted.

The evolutionist has a different view of science. He or she is happy to accept that evolution cannot be proved in the laboratory, for no one has watched a mollusc turn into a monkey, let alone a monkey into a man. The cynic may say that evolution is, to quote Bryan again, 'millions of guesses strung together', but the resulting web appears to map accurately the world in which we live. We cannot go out into the fields and watch a new species being formed. But we can observe natural selection at work in small ways and draw conclusions on the earth's history based upon fossil remains and geographical distribution of species today. And from these clues we can piece together a theory which appears to explain most, but never all, of what we see.

As Darwin wrote to a friend just over a year after *Origin* had been published,

'I am actually weary of telling people that I do not pretend to adduce direct evidence of one species changing into another, but I believe that this view is in the main correct, because so many phenomena can thus be grouped and explained.'

Here the starting point is human hypothesis, and this is tested against what is known in nature.

The same approach when applied to the Bible focuses on the humanity, rather than on the divine givenness, of the writings. Immersed in ancient culture, the message needs to be 'interpreted' for the underlying truths to become clear. The Bible is not a bald collection of statements which can be understood without reference to the context in which they were written. And we, too, can only perceive God using the thought-forms of our age. The Bible is a priceless library of such accounts.

With this viewpoint both science and Scripture are realms of interpretation. They start from man and highlight God and reality.

The two ways pull in opposite directions. At the extremes, the creationists move towards the givenness of Word and world, with man as passive recipient; their opponents start with human speculation and use this as a mirror which, through constant polishing, comes to reflect reality.

On this map the scientist and Christian believer may gravitate to either pole, and there plant his or her flag. People being people, some may even plant more than one flag, looking at the world in a different way on Sundays than they do the rest of the week! The intention here is not neatly to box people of differing persuasions, but to emphasize that the issue is not science versus faith, or even evolution versus creation.

It is about the way we understand reality.

---

subsequently split, with one of its offspring, the Institute for Creation Research, becoming the largest and most influential of the scores of such societies now dotted over the Western world. In Europe the movement never took off to the same extent, partly because of the more widespread acceptance of evolution by the churches and partly because the social factors in the United States were so different.

With the coming of creationist societies, the opposition to evolution was no longer focused in a few distinguished academics. The days of Louis Aggasiz had long gone. The age of specialist pressure groups had emerged, or should we say evolved? The trial of 1925 by no means defines the death of creationist opposition. Admittedly, with the establishment of the new synthesis, the doubters no longer debated in the scientific journals. By the centenary of the publication of *Origin*, Darwin's views were supported in a way never experienced in his lifetime. Yet creationist ideas continued to find widespread support among ordinary people. The appeal to common sense, to the uniqueness of mankind, to the importance of morals, ensured a following to ideas far out of step with scientific orthodoxy. When, in 1963, two Memphis State University teachers were reprimanded for discussing evolution in class, the now elderly John Scopes commented to a reporter:

> 'I think the case was lost in 1925. We fought a good battle but we didn't win very much. I would hope it would be different today, but I don't really know.'

Did the evolutionists win the intellectual battle in 1925? Forty years later Scopes was doubtful. Nothing is clearer than the changes in school textbooks. The scientists had not changed their minds, nor had the authors of textbooks. The current was running strongly in Darwinian riverbeds. Yet popular pressure dictated that instruction on evolution was kept to a minimum. In one major biological text the section on Darwin's life was reduced from 1,373 words to 45, and that on evolution from 2,750 words to a mere 296. The creationist societies also produced their own literature, some of it of very high standard and aimed at the high-school market.

The teaching of such societies centred around:

☐ *A short age for the earth*, of around 10,000 years rather than the commonly accepted 4,600 million years;

☐ *The creation of mankind according to a historical and literal reading of the early chapters of Genesis.* Groups vary as to whether or not the 'six days' are to be taken literally or as representing six time epochs. But the Bible account is the foundation of any scientific and historical explanation of biological origins;

☐ *The reality of the Flood* as a major factor in the geological formation of the earth and the deposition of fossil strata;

☐ *A rejection of evolution as 'fact'*. It is simply a theory, and public schools should allow the teaching of other explanations of human origins, namely the Genesis revelation. They believe the story of Genesis is upheld by their own rigorous scientific enquiry.

It is this last point — that Genesis should be taught in schools — that has caused the most controversy. In particular the debate reached the world headlines when an Arkansas law allowing equal time for the teaching of evolution and 'creation science' was tested in the courts.

Evolution in the classroom

Arkansas Governor Frank White signed Act 590 without even reading it. And the senator who had introduced the bill on 24 February 1981 had not written a single clause himself. Yet the bill had far-reaching implications for it allowed into the state schools the teaching of 'creation science' as an alternative to evolution. Said Senator Ben Allen:

> 'I voted for it, but I hadn't read it and didn't really know what it said. We discussed it for about a minute. The assembly did not pay attention. Most of us felt it was meaningless, just a piece of junk, so why not vote for it. I am sick as hell.'

The legal battle which ensued to revoke the bill was quickly dubbed 'The Monkey Trial 2'. People saw it as a re-run of Dayton, Tennessee. But since 1925 the focus of argument had shifted. Ostensibly the religious and moral purposes, so central for Bryan, were forgotten. Creation science was promoted by its supporters as a more accurate *scientific* account of the earth's origins. On trial was the right to give equal classroom time to this alternative theory. And this was backed by the apparently reasonable appeal that to limit scientific education to just one approach (namely that of evolution) was a travesty of fair-minded enquiry.

Whether or not the underlying religious motivation was indeed subsumed under a passion for fairer science will be tackled in a moment. From the start critics claimed the 'two theory' approach was a Trojan horse within the educational camp and that whatever the publicly stated intentions of the creation scientists, their hidden agenda was the restoration of Christian belief and morals. From the large church denominations there was also opposition to the new act. It was not because they feared religion! But they were wary of the approach behind the legislation.

As Federal District Judge William Overton sat down in court 419, Little Rock, Arkansas, he was confronted by a battery of almost a dozen lawyers from the American Civil Liberties Union. Their case was that 'creation science' was not really science, but religion. As religion it therefore infringed the first amendment of the United States constitution which bans the teaching of religion in schools. Many countries support religious teaching in state schools, but the United States has rigorously rejected such teaching as being divisive. Teaching religion, in the American system, is purely a matter for the churches.

As with the first Monkey Trial, the legal question was not the validity of evolution or six-day literalism. Darwin was not standing in the dock. The issue was whether the creation theory was really science, or a religious philosophy dressed up to look like science. The evolutionists put forward as witnesses a dazzling display of notable scientists, including the geneticist Francisco Ayala and the palaeontologist Stephen Jay Gould. By their choice of these two witnesses alone they laid to rest a common misunderstanding, that if you discredit Darwin you disprove evolution. For Ayala is a committed Darwinian, while Gould has his reservations about Darwin's emphasis on gradual change. As we have already seen, Gould believes evolution goes in jumps — a process he calls 'punctuated equilibria'. To his mind, the puzzling jumps in the fossil records are not gaps without data (which we hope one day to fill with transitional links), but reflections of periods of rapid change. Since evolution moved so fast during these years, we are unlikely to find intermediate forms.

The two men differed over nature's mechanisms and speed. Yet both firmly upheld evolution, and would have no truck with the miracles invoked by the creationists. In power of argument the evolutionists held the day.

Stephen Gould's testimony was stopped after only half an hour; he had made his point. As a fellow witness, Michael Ruse, amusingly records:

'I have never seen such a disappointed man ... Steve had been looking forward to a cosy afternoon putting us all right on the gaps in the fossil record, and he was finished before he had begun: rather than talking of gaps in the record, he was condemned forever to be one.'

The central debate was, however, not the validity of evolution, but the legality of teaching creation science. Was it, or was it not, religion? Judge Overton decided it was, and should be banned from state schools.

He found that in the writing of the new law, the creationists had indeed been motivated by religious concerns. This was clear from their correspondence, their books, and the societies to which they belonged. Under United States law this was incongruous with a commitment to keep religion out of the classroom. Other countries, faced with the same question but with fewer qualms about religion in schools, might rule differently.

The judge understood science as a procedure to derive testable laws from observations of natural phenomena. By definition, miracles are excluded. Creation scientists, he observed, make no attempt to conceal their appeal to miracles. Their argument is that we may observe today a whole series of natural phenomena, such as apples falling to the ground and atoms decaying at known rates. But at creation different processes were used by God, processes beyond our understanding or investigation. These are miracles. And it was these that Judge Overton ruled as inadmissible to be taught in state schools.

Whether or not teaching creationism is constitutional, it will for a long time remain popular. The trial indicated the widespread support for the Bill's enactment. But, as the Judge pointed out, constitutional government is not determined by majorities.

The religious dilemma

Judge Overton banned miracles from being taught in the classroom as science. A view of science that rules out miracles from the start may appear to cut right across a religious world-view. Christians have always believed in the resurrection of Christ, the supreme miracle, and cannot accept that God does not intervene in his creation. That, as they say, is the bottom line. Christianity must always admit the possibility of miracles but the crucial question is the degree to which they can be included in scientific explanations. Intellectually there is no embarrassment. For if God is behind all phenomena, and science only the record of his activity, then a miracle is an occasional departure from his regular way of working.

For practical purposes any scientific discipline needs to limit itself to so-called natural phenomena. It can deal only with the normal patterns of events. Undertakers do not expect the deceased to rise, nor do scientists expect nature to depart from the rule of cause and effect. A too-ready recourse to explanation by miracles stunts scientific enquiry. Instead of saying 'We do not yet know, we must keep investigating', much creationist literature turns quickly to a special miracle in order to hold the pieces together.

In an increasingly secular world there is a temptation to reassert God's existence by seeking him in the miraculous. A God who intervenes through miracles is felt to be more evident, more powerful, than one who works only through the regular laws of nature. To demonstrate the reality of God we demand miraculous signs from him. Dangerously, theology slides into idolatry, as we create God in our own image to bolster our lack of

---

A MATTER OF WHAT WE MEAN BY SCIENCE

At the Arkansas 'Monkey Trial' the judge, William R. Overton, stated that the essential characteristics of science were:

☐ *It is guided by natural law, and uses natural law to explain phenomena;*

☐ *It is testable against evidence and events found in the empirical world;*

☐ *Its conclusions are tentative, that is, they are not necessarily final;*

☐ *It is falsifiable, that is future evidence may be found which destroys the theory.*

Against this definition of science, 'creation science' does not measure up. Creation scientists readily refer to the miraculous power of God in order to explain what they see as otherwise inexplicable events.

'We do not know how God created, what processes He used, for God used processes which are not now operating anywhere in the natural *universe*. This is why we refer to divine creation as Special Creation. We cannot discover by scientific investigation anything about the creative processes used by God.'
D. Gish *Evolution? The Fossils Say No!*

In other words, they refuse to limit their understanding of the earth's origins to simply natural causes. Furthermore, they hold the Scriptures as the final arbiter on what did or did not happen at creation. Scientific evidence such as rock formations, fossil remains, and animal structures, must in the end be used to uphold the voice of Scripture. The Bible must have the final word. Creation science is testable (ultimately) against the Genesis account, not the empirical world.

That creation science fails to meet the above four-fold test is no embarrassment to its promoters. They claim the evolutionary account of the world also fails to meet this stringent definition. If creation science is guilty then so, too, is evolution science. Yes, it espouses natural law and decries miracles, but as a theory it cannot be tested in the laboratory, and seems impervious to challenge. We simply cannot go back billions of years to observe or test our ideas. And, say the creationists, serious scientific criticisms of evolution are simply brushed aside. Where then is its vaunted claim to be testable and falsifiable?

This, then, is the method of the creationists. Rather than elevate their 'creation science' to the status of a scientific theory alongside evolution, they try to knock evolution from its science pedestal. Both, they say, are only ways of picturing creation. They are 'models' and not like the demonstrable and testable law of gravity. As models they are ways of understanding the past, and a Christian has as much right to a view of the past which sees God at work as the secular humanist has to leaving miracles out altogether. If

faith. The issue is not whether we *prefer* a God of a miraculous six-day creation, but whether in fact God has operated in this way.

The Victorians pleaded that design in nature was proof-positive of a Grand Designer. When Darwin showed that the same result could be obtained by the action of regular laws it threatened the belief of many. They failed to see God working even through Darwin's laws. In a similar way today the creationist insistence on a literal reading of Genesis and a short age for the earth may bolster the faith of some, while destroying Christian credibility for many. Others have no difficulty in seeing God at work in the regular and commonplace. A recent commentator on Darwin's work, a professor of physics, has remarked, 'We have Darwin to thank for finally making it clear that God is not a secondary cause operating on the same level as natural forces, or a means for filling gaps in the scientific account.' You can almost sense the relief in his voice.

A cartoon by Heath in the British newspaper 'The Sunday Times' reflects recurrent controversies over evolution which have hit the news time and again through the past century and more.

'creation science' is religion then so is evolution. Both adopt a particular view of how things are. One sees God working through miracle and regular phenomena, the other sees only blind chance and natural law.

In his judgment Overton was not persuaded by this creationists' debunking of evolution.

Is evolution falsifiable? The court heard how the discovery of well-developed fossils in the earliest rock strata would destroy evolutionary theory; a point Darwin himself had recognized.

Is evolution impervious to attack? Here the evolutionists dismissed the criticisms of the creationists because they were not substantial. To use the crossword analogy again, most of the scientific clues fit. An occasional scientific fact out of place (or letter in our analogy) does not justify redoing the whole thing.

Judge Overton took as his starting point the definition of science given above, and showed that creationists appealed to miracles and the witness of Scripture. He referred to Act 590's requirement to teach the alternative model of sudden creation from nothing:

'Such a concept is not science because it depends upon a supernatural intervention which is not guided by natural law. It is not explanatory by reference to natural law, is not testable and is not falsifiable.'

In a previous age the essence of science was to discover God's ways of working. Miraculous interventions were perhaps rare, but certainly permissible. They would have found Overton's dismissal of miracles presumptuous. They believed that if the task of a historical science such as geology was to understand what had happened in the past then it would be a rash judgment to exclude the possibility of the Creator becoming involved through miracles. The scientist should employ wise caution, not begin with outright rejection.

But, as we have seen, in our secular world we have redefined what we understand by 'science'.

Who has the right definition?

IGNORANCE OR INVESTIGATION?

In the nineteenth century Charles Lyell feared that the too-ready recourse to miraculous explanations held up the progress of science. The same fear is prevalent today.

To take one example. J.C. Whitcomb, co-author of the famous *Genesis Flood*, has discussed in a subsequent book the tricky problem of how the different animals each made their way to Noah's ark. He writes:

'When all else fails, why not cut the Gordian knot of endless speculation and simply acknowledge that God, and God alone, had the power to bring two of each of the basic kinds of air-breathing creatures to the Ark?'

Why not indeed? The science fiction writer Isaac Asimov retorts:

'With creation in the saddle, American science will wither. We will raise a generation of ignoramuses. We will inevitably recede into the backwaters of civilization.'

And from the previous century Thomas Huxley joins in the attack:

'The hypothesis of special creation is, in my judgment, a mere specious mask for our ignorance.'

The British Museum of Natural History, in South Kensington, London, caused some controversy in 1981 when it mounted exhibitions which avoided issues of evolutionary descent.

A religious understanding of history (involving the resurrection), and a religious understanding of science (including creation) needs to proceed with care. To rule out from the start the possibility of God's intervention is to prejudge the issue. But to use God as a fill-in actor for our own ignorance does no credit to the minds he has given us. Many scientists with religious faith prefer to see God working behind the natural phenomena they study. They do not doubt the reality of God. They understand scientific 'laws' as reflections of his constant activity. Only on exceptional grounds will they declare

something a miracle. For them the creationists have overstepped the mark.

And yet there is a danger round the corner. Scots theologian T.F. Torrance is concerned that by thinking in terms of this world's chain of cause and effect we are gradually blinded to the other, divine, world. As he expresses it:

'The problem of natural science . . . is that in developing autonomous modes of scientific investigation it is tempted to treat the universe as a self-sufficient necessary system, which does not need to be understood by reference outside of or beyond it.'

Creationism warrants attention. It is a backlash against a secular creed. Behind the scientific fight lies a determination to restore a world-view that includes a spiritual dimension, and to advocate a secure morality based on unchanging truths. Those who fight for it are convinced that evolution is not just another theory. Rather, it promotes an antipathy towards the things of God and imparts values that are centred on ourselves. 'Descended from apes' implies a brutish society; 'evolved over aeons' suggests an absence of God.

'Evolution' is not just about the origin of mankind, but about the development of his pride.

9
MELTING DOWN
THE GODS

Our understanding of biological origins is always changing, always evolving. In a previous chapter we saw how the rediscovery of Mendel's laws led to an emphasis on large mutations as the source of change. Only when it was realized that our bodies, our height or the shape of our nose, are the products of many gene changes working in concert did the role of mutations begin to take a back seat. They came to be seen as the well of variations on which the power of natural selection could draw.

Throughout the history told in the previous chapters, the biological theories have always given rise to philosophical ideas — explanations as to the meaning and value of life which have emerged due to the stimulus of the scientific advances. But sometimes the philosophies have moved into first place, and deeply held beliefs have determined beforehand what is scientifically to be believed. The dog takes its master for a walk.

Yet can we ever call one master to the other? Is it not more a marriage relationship, with values prejudicing perceptions, and scientific 'facts' influencing philosophies? The story is not of religion somehow at war with science, but of a matrix of scientific theories vying for predominance amid clashing social and religious outlooks.

The rise of creationism has been treated almost in isolation, but the reality of course is more complex. If evolution was a benign theory that affected no one except biologists, then in the America of the 1920s the rhetoric of William J. Bryan could have been reserved for more political affairs; the school textbooks

of the 1970s could have been left undisturbed. But as secularism spread across the Western world, evolution itself became the focus of a twentieth-century myth. Mankind was believed to be progressing, ever moving upward. It became a religion in itself. And nothing more clearly elucidates this corruption of evolutionary science into a passion for progress than to trace the path from Thomas Huxley, bulldog of Charles Darwin, to Julian Huxley his grandson, founder of the religion of evolutionary humanism.

From Huxley to Huxley

Thomas Huxley was born in 1825. Like his friend Darwin, he rose to scientific eminence through a world voyage. In the twentieth century the Huxley family was to produce Julian, a biologist and popularizer of science, Aldous, a famous writer, and Andrew, a physiologist who received the Nobel Prize in 1963 for work on nerve systems. But Thomas was their first 'success'. Born into a family of five other children he attended an ordinary school for only two years, and was largely self-taught. At the age of twelve he was reading Hutton's *Geology* in bed by candlelight. Through sheer dint of effort he won prizes in medicine while still a student. Then, accepting an offer from the Royal Navy, he sailed as assistant surgeon on HMS *Rattlesnake*, making a round trip to Australia and New Guinea.

Soon after his arrival back in London in 1850, Huxley met Darwin and the two became friends. We have already discussed his defence of *Origin of Species* in the (now) famous clash with Bishop Wilberforce. This was his first main adventure in public debate, and the success of his gifted speaking persuaded Huxley to 'carefully cultivate it and try to leave off hating it'. Thereafter he became the public champion of science, taking on opponents through the pages of the Victorian periodicals with an acerbic style and a passion for scientific truth. A more cynical view sees Huxley's attacks on the church as a struggle to wrench cultural leadership away from the clergy to people of science like himself.

But a fighter he certainly was. His statue stands alongside Darwin's in London's Natural History Museum. It captures the difference between the two men. Both are sitting, but whereas Darwin has his hands clasped passively in his lap, Huxley holds one fist clenched.

'I profess myself the most debonair of dogs; but I have been trained to follow at the heel of true science, and I cannot undertake not to bark, perhaps even to bite, whenever I observe that someone, who is not my mistress, is assuming her authority or trying to wheedle me from her side.'

He became known as Darwin's bulldog, but in reality his true allegiance was not to the particular theories of his friend, but to Darwin's scientific method. Huxley was determined that science should not be plagued by the constant appeal to divine or metaphysical causes. He warmed to Darwin because he wrote of nothing but natural cause and effect. He had his own criticism of natural selection, and on more than one occasion dismayed his friend by his failure to promote the ideas of evolutionary change. Huxley set the highest importance not on the validity of one theory or the next, but on an approach which brought closer the goal of expressing all phenomena in terms of scientific laws.

In an 1860 review of *Origin* he wrote of the immaturity of creationist views, and the wonder of a universe held together by immutable laws:

'The hypothesis of special creation is not only a mere specious mask for our ignorance; its existence in Biology marks the youth and imperfection of the science. For what is the history of every science but the history of the elimination of the notion of creative, or other interferences, with the natural order of the phenomena?

'Harmonious order governing eternally continuous progress — the web and woof of matter and force interweaving by slow degrees, without a broken thread, that veil which lies between us and the Infinite — that universe which alone we know or can know; such is the picture which science draws of the world.'

Although Huxley speaks here of the Infinite, he was no Christian believer. His was an infinity veiled from human sight by the world of interlocking forces and laws. Only knowledge of the natural world was possible. For the rest, he remained 'agnostic' — an apt description since he was the inventor of the word.

But though he was no believer and offered no prayers to a god, he was deeply religious. His vision of all-permeating natural law was a spiritual resource in his life, satisfying in Huxley deep religious longings. He wanted to sit down before fact 'as a little child' and humbly learn. Though he bitterly attacked the Victorian church for its scientific ignorance and superstition, his own feelings when faced with the beautiful order of the cosmos rivalled in religious intensity those of the most ardent churchmen. He believed in 'cherishing the noblest and most human of man's emotions, by worship "for the most part of the silent sort" at the altar of the Unknown'.

Huxley's god was Natural Order. He abhorred all notions of chance. Everything — even humanity's own nature — functioned according to law. And the doctrine of providence thus formed seemed to him 'far more important than all the theorems of speculative theology'. Nature no longer led up to nature's God, but basked in its own glory. It was sufficient for mankind

Thomas and Julian Huxley, grandfather and grandson, both believed that science provides the key to understanding life as a whole.

to understand the ways of science. Why should we probe into the unknown or, rather, into the unknowable? Thomas called himself a 'bishop' and the worshippers at the altar of naturalism the 'church scientific'.

There is a touching photograph of Thomas with his grandchild Julian, and there was evidently a great deal of affection between the two of them. He believed that Julian liked biology, and expressed to his son (Leonard Huxley — Julian's father) his desire to train the boy. Although Thomas died in 1895 in one way his wish was fulfilled, for the young scholar was educated at Eton using scientific laboratories built by the determination of his grandfather and locally known as 'Huxley's Folly'.

Julian went on to Oxford University where he gained a first-class honours degree in zoology, as well as the Newdigate Prize for poetry. In his autobiographical *Memories* he wrote that, 'looking back, I seem to have been possessed by a demon, driving me into every sort of activity'. He researched and he lectured — both in England and America. He made significant contributions to the growing new synthesis in evolutionary thought, and because of his concern for education and the Third World he was appointed the first director-general of UNESCO. He was one of the most influential popularizers of science in his age, joining with H.G. Wells to write a monumental work, *The Science of Life.*

Julian Huxley imbibed not only his grandfather's love for biology but his sense of wonder before nature and his agnosticism towards God. In fact the favourite grandson went even further. He was not merely agnostic but believed 'that the idea of personality in God has been put there by man'.

This outlook, he said, stemmed from his biology. On the long progressive path of evolution, mankind has used religious concepts to understand nature: to express our sense of the sacred, and to undergird our schemes of morality. But now, with the advent of scientific knowledge, these old forms of expression are out-dated. Religion was only ever a human language used to express our deepest feelings. But now it is a dead language, and needs to be discarded from twentieth-century conversation. Now we can see clearly, and we can look reality in the face:

Julian Huxley was instrumental in developing a school of thought known as 'evolutionary humanism'.

EVOLUTIONARY ETHICS

As we move across the generations between the two Huxleys there is also a shift in approach to morality. With their interest in evolution both men examined with care the impact of biology on ethics. Was it possible to derive a system of ethics from evolution? The comparison is made all the more easy since both men delivered lectures on the subject sponsored by the same Romanes foundation. Thomas Huxley spoke in 1893, and his grandson Julian in 1943.

Thomas Huxley firmly answered 'no'. We cannot derive ethics from evolution or creeds from the cosmic process. The understanding of natural selection commonly used in Victorian times, nature 'red in tooth and claw', did not allow him to use evolution as a model or example for human behaviour. Society needs co-operation, not competition. If there was to be a struggle, then it must be within the human heart trying to rise above the baser instincts inherited from an animal past. Here Huxley was faced with a paradox. He was adamant that our habits, our emotions, our social systems were all formed through our evolutionary history. And yet when it comes to morality we can somehow rise above this all-determining past, and what animal instincts remain we must quickly dismiss. As one reporter who listened to Huxley's lecture commented:

'If evolution means anything, the ethical process must be as strictly natural as the cosmical . . . Professor Huxley came dangerously near the edge of the old dualism of nature and the supernatural, if he did not fall over the cliff.'

Fifty years later Julian Huxley took the opposite view. He believed that the history of our biological origins could also teach us how to behave in the present. He was less troubled than his grandfather over the competitive element within natural selection. The Victorians were wrong in their emphasis. Species evolve, not by killing the weakest among them, but by the strongest being more likely to reproduce. And since evolution is generally reckoned to have 'progressed' (for example, we are higher organic forms than the amoeba or the monkey), then the overall outcome is good. So we must get in step with evolution if we want this improvement to continue. What is right and moral is therefore anything that fosters evolution, and anything that respects human individuality and development.

We have already remarked on the danger of deriving values from biology. Julian Huxley argues in a circle. He defines what he means by progress and then uses the fact that it has happened as good reason for us to promote it in the future. For his commitment to human individuality

'The ultimate task will be to melt down the gods and magic and all supernatural entities.'

Yet we still need to preserve the sense of the sacred, and so the role of religious explanations must be taken over by a similar system, this time constructed on the basis of science. Humanity has always needed, and will always need, some overarching philosophy around which life can be focused. Huxley wrote in *The Humanist Frame*, 1961:

'If the situation is not to lead to chaos, despair or escapism, man must reunify his life within the framework of a satisfactory idea-system. He needs to use his best efforts of knowledge and imagination to build a system of thought and belief which will provide both a supporting framework for his present existence, an ultimate or ideal goal for his future development as a species, and a guide and directive for practical action. This new idea-system, whose birth we of the mid-twentieth century are witnessing, I shall simply call "humanism".

'It must be focused on man . . . It must be organized round the facts and ideas of evolution . . . It will have nothing to do with Absolutes, including absolute truth, absolute morality, absolute perfection and absolute authority.'

Of course no temple was ever built, or hymnbook published. If people gathered to worship it was in the scientific lecture room or in front of televisions watching the latest nature programme. Humanism's sense of the sacred was derived from the wonder of nature, and its unifying hope from the onward march of evolution. Whereas the elder Huxley had seen the interlocking web of natural law as

he reads back into biology his own liberal democratic tradition. But this is somewhat at variance with what we would learn from, say, a beehive where the life of an individual worker is of no consequence at all. Studying a colony of baboons might recommend a hierarchical society. Who is to decide which biological analogies are the right ones to use? He (rightly) condemned Hitler's use of force, but the Third Reich was supposedly based on the biological values of the strongest survives. Huxley criticizes their biology, and he is correct, but it seems suspicious that the values he happens to derive from biology so clearly match his own British tradition.

His grandfather rejected evolutionary ethics because his own values were at variance with the then current biological theories. Julian Huxley embraced evolutionary ethics because the new twentieth-century biology had by then minimized the role of competition.

In the end we are left with the advocacy of change. This seems to be the lasting legacy of an evolutionary viewpoint: change is to be encouraged, and fixed forms of authority or morality despised. As Julian Huxley says, there are to be no absolutes. Except, of course, the absolute good of change itself. It may be pragmatic, but who is to say that this is *good*?

a net which engulfed humanity, dragging him wherever it would, Julian believed that human consciousness now gave mankind the power to direct the further course of evolution. He saw that his grandfather had 'put evolution on the map' and that his task was now to 'map evolution'. By this he meant more than simply exploring the details of the science which were unknown to Darwin and his friends in the last century. Huxley meant that we should explore ways to direct the future course of evolution, drawing the map of life for ourselves. So Thomas Huxley's 'Providence' was changed to Julian Huxley's 'Progress', and the worship of order within nature was transfigured into the goddesses of change and human development, with humanity controlling the shape of the world's future.

As an organized belief-system humanism never really took off. Humanist societies sprung up around the Western world, but they never attracted the mass of ordinary people. Yet basic humanist ideas did take hold at an important opinion-forming level. In one popular science book after another this awe in the face of creation, this establishing of humanity by evolution, was (and is) trotted out.

There is now a general underlying approach which informs much of Western thought. It holds that:

☐ *Evolution gives humanity a place and a role in cosmic history;*

☐ *Everything, including our morality, is subject to evolutionary development, and therefore no longer fixed;*

☐ *Eventually by our own wisdom we will progress higher and higher.*

All these have become a common currency of belief. And yet exchange for the former coinage of God and prayer has left little in the way of savings for use in times of personal suffering, and even death.

More than a decade after Julian Huxley's own death in 1975 we may question his grand optimism. We have seen wars to end all wars; we live under the threat of nuclear annihilation; we pollute our environment; rich nations refuse to share wealth with poorer nations. The signs of human moral progress are few and far between. The popular science-writers know all this, and they explain mankind's ignoble past. Yet they still hold out a glorious future. They still believe we will rise above our animal origins. They must do . . . for otherwise they have a bleak religion indeed.

The eighth-century Hebrew prophet Isaiah described the coming rule of God as creating 'rivers in the deserts'. He spoke of God bringing healing, peace, and a deeper knowledge of himself. Richard Leakey, the famous anthro-

TEILHARD DE CHARDIN: THE OMEGA POINT

Pierre Teilhard de Chardin was born near the French town of Clermont-Ferrand in 1881. His mother was the great-grandniece of Voltaire, but her child was no biting critic of the established church as was that famous philosopher. Or so it seems in retrospect, although during his lifetime the Roman Catholic Church banned some of Pierre's scientific appointments and sent him to the relative backwaters of China to mend his ways. He was seen as a dangerous innovator, and he was forbidden to publish his works.

What had Teilhard de Chardin said that seemed so shocking? The young Pierre joined the Jesuits in 1899, and when qualified began to teach in an Egyptian secondary school belonging to the Society. It was in the Egyptian desert that he began his lifelong study of minerals and fossils. He was ordained and entered the Sorbonne to study geology and palaeontology. The First World War interrupted his studies, and it was not until 1922 that they were completed

Teilhard de Chardin attempted to find a new and modern way to present Christian truth making wide use of ideas drawn from evolution.

and he was offered the chair of geology at the Institut Catholique in Paris. It was here the trouble started.

As a committed scientist and a devout Christian, Teilhard de Chardin was determined to find a viewpoint which harmonized both of his great loves. He was convinced that 'a reconciliation must be possible between cosmic love of the world and the heavenly love of God . . . between the cult of progress and the passion for the glory of God'. As many believing scientists before him had acknowledged, the plans of God seen in his material creation and in his revelation of himself in Christ must be in accord. There is, after all, only one God. The scientist Pierre enthusiastically embraced evolution as the perspective from which to view the world. His

fossil studies pointed clearly to an ever-increasing complexity in animal types as the world spun on its axis, aeon after aeon. He was gripped, as was Julian Huxley, by the increasing level of consciousness: the arrival of human culture or, as Teilhard expressed it, the 'noosphere'. Whereas Huxley melted down the gods and sat humanity in the driving seat of evolution, Teilhard believed this progressive development was the work of God.

'A creation of evolutionary type (God making things make themselves) has for long seemed to some great minds the most beautiful form imaginable in which God could act in his universe.'

Teilhard's vision was to use the evolution of the past as a prophecy for the future. Looking back to how life began he noted how the molecules had linked together to form more complex combinations. So now the future held out the hope of greater linking together among isolated human beings, with ever greater sharing of thought and consciousness. Even today we speak of the world being a 'global village', yet Teilhard was looking way into the future. To plot that future is, of course, difficult, just as an amoeba could not have foreseen an American city! Yet if the drive is towards increasing complexity, and yet coherence, Teilhard was confident that we would reach a point when men and women would combine with

pologist, transmutes these very same qualities on to his god of science — except that the only increase in knowledge will be of ourselves. He writes, in his book *Origins*:

'The potential is enormous, almost infinite. We can, if we so choose, do virtually anything: arid lands will become fertile; terrible diseases will be cured by genetic engineering; touring other planets will become routine; we may even come to understand how the human mind works! What is at issue is whether nations can live peaceably with nations and with an understanding and deep respect for the natural world they inhabit so that one day these, and other, predictions may be fulfilled. The answer, emphatically, is Yes... We are One People, and we can all strive for one aim: the peaceful and equitable survival of humanity.'

one another to form a kind of super-organism in which a communal consciousness, a suprapersonal unity, would be manifest. He called it the Omega Point.

Teilhard related this cosmic vision to his Christian faith. He saw in the historical Jesus Christ not only the perfect man in history but also a sign of the future. And he saw in the cosmic Christ the source of the unifying energy that would bring human history to the Omega Point. Teilhard stressed those parts of the New Testament which speak of all creation being brought to unity in Christ: Christ himself is the Omega Point.

Now in stressing this cosmic role of Christ, with the world progressively evolving, his thought appeared to underplay the place of failure and sin. Traditional Western theology sees the world more in need of restoration by Christ than on target for a glorious future. There is an emphasis on how much we have missed the mark of what we were intended to be. We have fallen from a state of perfection, not slowly ascended a path of advancement. Whatever the future (and all Christians expect the final summation of history to be accomplished in and by Christ) we cannot ignore our present imperfection.

Teilhard himself was not unaware of the reality of sin. He firmly believed that mankind and the world were not perfect, yet in his writings he anticipated the future rather than regretted the past. And it was these writings which caused him difficulties. Because of a paper he wrote on the relation between evolution and the doctrine of original sin he was removed from his position by his Jesuit superior.

From France he went to China, where for twenty years he researched with fellow Jesuit scientists the rich fossil deposits around Peking and Tientsin. He became a leading authority on ancient man, and on his return to Paris after the Second World War he became director of research at the Centre National de la Recherche Scientifique. Even then his views blocked further advancement, and his superior-general refused permission for him to take the chair of palaeontology at the Collège de France. Teilhard was out of step with the official Vatican line on evolution. A few years later he moved to New York and continued his research from there.

His writings still remained unpublished, and were only printed following his death on Easter Day, 1955. The fifteen volumes, many now available in popular editions, show a man of great originality, with a passionate concern to integrate the worlds of science and belief. Many have been enthralled and uplifted by his vision of humanity moving towards its Omega Point. Many now share with Teilhard a belief that evolutionary progress lies at the very centre of God's plan.

The Vatican's understanding of evolution has broadened. In 1943 Pope Pius XII published an encyclical, *Divino Afflante Spiritu*, which acknowledged the enormous progress made in biblical studies since the previous century. This untied the straitjacket of literalism which bound the interpretation of Genesis, and in his 1951 encyclical *Humani Generis* he also allowed discussions on evolution between scientists and theologians. The Second Vatican Council pronouncements of the 1960s saw a further accommodation to modern thought, and more recently theologians such as Karl Rahner have been able to expound theology with full recognition of science's understanding of humanity based on evolution and anthropology.

Teilhard's own thought is speculative, and he has been criticized by many for mixing scientific description with metaphysical fancies. He may have believed in the inevitability of progress, but others find this wholesale extrapolation of the past as a means to predicting the future a very shaky foundation indeed.

Nevertheless Teilhard was right to focus on evolution as the new perspective of the twentieth century. He was also realistic in believing that the church needs to express its deeper truth concerning human destiny within that framework.

Julian Huxley, along with others, established evolution as a way of looking at the world and of finding humanity's place within it. It is no longer just a biological theory: it is a symbol, a comforting dogma, an imperative for action. It holds out a future and it heals our wounds:

'Evolutionary truth frees us from subservient fear of the unknown and supernatural . . . It shows us our duty and destiny . . . It gives us potent incentive for fulfilling our evolutionary role in the long future of the planet.'

FALLEN HUMANITY

Christian thought has always been unhappy with an emphasis on progress, especially when the source of advancement is seen as humanity itself.

Christian belief understands humanity as being created in the image of God, and yet at the same time capable of sickening depravity. We are saints and sinners. Acts of true altruism took place within the horrors of Auschwitz. We possess the capacity for good, yet the freedom and ready inclination to serve our own selfishness. In speaking of 'sin' the New Testament refers not so much to particular deeds and faults, as to the degree we fall short of what we were intended by God to be. Its most common word for sin comes from archery — it means to miss the mark.

Traditional understanding of the first chapters of Genesis sees the origin of evil in the world as stemming from Adam and Eve's disobedience. It is from their original transgression that all future generations have been tainted.

How should we regard this story today? The creationists believe in a historical couple, and a literal 'fall' at the outset of human history. Genesis is to be taken literally and, say the influential writers Whitcomb and Morris, 'this is absolutely essential to the entire edifice of Christian theology, and there can simply be

no true Christianity without it'. The creationist inserts into his reading glasses lenses that only focus on the literal meaning of Genesis. The God revealed through scientific endeavour is only glimpsed out of the corner of an eye.

Others find this shortsighted. They see that the vast majority of scientists support evolution. They argue that since 'Adam' is simply Hebrew for 'man', the fall of Adam and Eve may represent the disobedience of the race. There may have been an original couple, somehow different from the other evolved hominids, on whom God bestowed his Spirit and who subsequently rebelled.

Christian biochemist Tim Hawthorne has written:

'Most conservative theologians believe that the whole human race descended from that original pair. Others see Adam (the Hebrew word for *man*) and Eve (Hebrew, *living*) as representing all the humans of their day, Genesis itself indicating that there were others, such as those among whom Cain went to live. Whether or not Adam and Eve were the only human pair alive at that time, they acted representatively on behalf of all mankind. So as I see it, the doctrine of the fall of man does not require us to believe that a single human pair existed *as the only human beings* and we can regard Adam as the "federal" head of humanity.'

For many this is still to impose on the Bible a fondness for literalism and time-scales that is foreign to the meaning of the text. They argue that to be fair to these chapters of Genesis we must acknowledge their

own particular style as literature. Undoubtedly the first chapters of the Bible are couched in the language of story — after this, then that. And so a perfect creation is followed by man's disobedience and the fracturing of relationships. But should we seek to date this moment of disobedience so many years BC and call Adam the New Stone Age man? Or is the story of Adam and Eve a literary device used to explain deeper issues? In other words, are we dealing with historical truth here, or is this story a magnificent, God-given poem expressing eternal truths about God and his creation?

Liberal scholars would remove all sense of Genesis dealing with the past. Just as these chapters, they say, are about the deep truth that God *is* (and not just *was*) creator, so 'original' sin concerns mankind's individual and corporate failure *today*. The truth revealed is existential, not historical.

Alan Richardson has expressed this view:

'The time element . . . must be discounted; it is not that *once* (in 4004 BC — or 100 thousand years ago) God created man perfect and then he fell from grace. God is eternally Creator. And just as creation is an eternal activity, so the 'fall' is an ingredient of every moment of human life; man is at every moment 'falling', putting himself in the centre, rebelling against the will of God.'

Which is correct? Each view claims to do justice to their understanding of science and the revelation of the Bible. It is the interpretative framework applied to Genesis that differs, the first two being more literal than the last, with Dr Richardson

Note how closely this mirrors the Genesis account. If one of the aims of Genesis was to assure the people of Israel that they need not fear the capricious (and yet fictitious) gods of neighbouring nations, so in the twentieth century 'science' has adopted this responsibility. With science to lighten our darkness today, we need not fear the perils and dangers which existed only in the night-time of our spiritual ignorance.

But where Genesis claims to be a revelation

even removing any sense of historical interpretation.

Whether Richardson is right to remove the fall from history entirely is doubtful. After all, chapters one and two of Genesis stand at the head of a book of Jewish history, not wisdom. But conversely we cannot force the ancient writer to follow our own twentieth-century canons of historical writing. These chapters are reflections on the relationship of God to humanity and the rest of the created order, and on the significance of evil in the world; they are not about details of its dating,

How can we explain the evil that repeatedly emerges in human history? The suffering of the Vietnam war was just one recent example.

whether at 4004BC or 4 million BC. The writer of Genesis knew that evil sprang from the heart of mankind (and not from some mysterious force of evil in the universe) and he describes how and why the world was shaped to accommodate fallen humanity.

God created time along with matter. Any revelation to us about the creation is bound to be expressed in language that uses concepts of time, even though God's actions stand outside time. Perhaps, then, these early chapters do not attempt to give us a strict historical sequence of creation and mankind's fall. They may be a witness to the creating of the sort of world necessary for human development *because* we are imperfect.

The writing is historical, rather than simply about humanity's condition now, in the sense that God *has* created the world in this way and we must understand the reasons for its, and our, imperfections compared to the glory that might have been.

On one thing all these Christian writers would agree. They are all certain of the reality of sin. And they would all be implacably opposed to the humanism of Sir Julian Huxley which focuses the hope of redemption on ourselves and not on the power and love of God.

THE MAKING
OF MANKIND

Richard Leakey is an anthropologist who has spent many years in Africa searching for fossils which might link us to our evolutionary past. It is a fascinating story with occasional finds dramatically interspersing long months of hard work. And as Leakey has studied mankind's past he has become aware of the fragile nature of our existence. Other species once dominated the earth, and may yet do so again. He points to the destructive potential of our atomic bombs yet feels that we have the capacity to rise above the need for warfare. He quotes Dobhansky as saying that we need a faith or hope on which to focus, and he ends his book *The Making of Mankind* with his own humanistic creed. His understanding of the past moves on to an ethic for the future. His science does not provide detailed moral rules, but it does (so Leakey believes) provide the hope we so desperately need.

Richard Leakey, who has worked for years hunting ancient human remains in the Great Rift Valley of East Africa, has done much to popularize ideas about human origins.

'For millennia man sought that faith outside himself, in a certain view of the world provided by religion. This position has been greatly eroded for many by the advance of science, particularly by the Copernican and Darwinian revolutions. I believe that there is a great deal of strength to be discovered by looking inside ourselves, with the knowledge that each of us belongs to the same highly successful and diverse species . . .

'Unlike our forebears who became extinct, we are an animal capable of almost limitless choice. The problem facing us today is our inability to recognize the fact that we *are* able to choose our future. Many people are content to leave their future to the will of God, but I believe this is a dangerous philosophy if it avoids the issue of our responsibility. It is my conviction that our future as a species is in our hands and ours only: I would remind those who rely on God's mercy and wisdom of the old adage "God helps those who help themselves". We must see the danger and the problems and so chart a course that will ensure our continued survival.

'For me the search for our ancestors has provided a source of hope. We share our heritage and we share our future. With an unparalleled ability to choose our destiny, I know that global catastrophe of our own hands is not inevitable.

'The choice is ours.'

of God, Huxley's scheme is a purely human construction. It is the creation myth of our day.

A chancy business

In nature Thomas Huxley believed he saw all-embracing natural law, and his favourite grandson unmasked the glory of progressive evolution. But others could only see luck.

During the 1950s Watson and Crick revealed the mechanism whereby one cell could pass on its information to a new cell. The DNA structure acts as a plan, and new genetic material is usually a faithful copy of the source DNA. Variations between a human child and its parents occur because the genetic code

inherited by the child is a *combination* of genes from both its parents. In human beings no two sets of fingerprints are the same. With sexual reproduction there is a shuffling of genes from one generation to the next.

Errors may occur in the replication of the genetic code itself. Large changes usually lead to deformities or diseases fatal to the plant or animal concerned. But small copying errors, or mutations, can be the source of advantageous variations. These natural selection grooms as the possible beginnings of a new species.

Whether they arise through gene shuffling or through gene mutation, evolution through natural selection depends on the production of random, yet minute, variations. But does this make life itself just a matter of chance? Or are the selections somehow determined beforehand, forming a stream of events that necessarily flows under a sort of evolutionary gravity?

Although Darwin had no knowledge of DNA he did realize that variations are random. And he confessed he was bothered by the problem. He did not want to say, as *Vestiges* had done, that God was directing the evolution of species. He wanted to remove the purposes of God as a direct scientific explanation. Yet the prospect of everything just arriving by accident horrified him more. He was troubled by 'the extreme difficulty or rather impossibility of conceiving this immense and wonderful universe . . . as the result of blind chance or necessity'.

Chance or necessity. Do we live in a world which is dependent on a roll of biological dice, or in one that is planned? Christian philosophers have always chosen the latter. They see the natural world as the outworking of God. The alternative creed, that we are here by chance, has been faced by a few atheists. In the 1920s Bertrand Russell wrote *A Free Man's Worship*, in which he bravely preached:

The pollution of air, sea and rivers by industrial effluent: an example of the flip side of human progress. Evolutionary humanism's optimism about humanity does not always seem to match the evidence.

Philosopher Bertrand Russell taught that it is harmful to try to explain the universe by using ideas not drawn from within the universe itself.

'That Man is the product of causes which had no prevision of the end they were achieving; that his origin, his growth, his hopes and fears, his loves and beliefs, are but the outcome of accidental collocations of atoms . . . that all the labours of the ages, all the devotion, all the inspiration, all the noonday brightness of human genius, are destined to extinction in the vast death of the solar system, and that the whole temple of Man's achievement must inevitably be buried beneath the debris of a universe in ruins — all these things, if not quite beyond dispute, are yet so nearly certain that no philosophy which rejects them can hope to stand. Only within the scaffolding of these truths, only on the firm foundation of unyielding despair, can the soul's habitation henceforth be safely built.'

Is this really the very scaffolding of truth, or is it the hangman's scaffold for human hope? Before considering these mighty alternatives we must come to some understanding of what we mean by 'chance'.

Suppose you are taking part in a card game, whether it be a rubber of Bridge or a simple game of Snap. You depend on the dealer not to cheat in shuffling the cards, and you watch that they are shuffled thoroughly. To remove any possibility of fraud perhaps you rely on some mechanical gadget to shuffle them automatically. That way you can eliminate human intervention and the cards dealt to you are random. You cannot predict what hand you will receive.

Or can you? If you designed the 'auto-shuffler' yourself then perhaps by advanced physics and abstruse mathematics you might predict the outcome. The shuffling is subject to the usual mechanical laws of the universe, as it would be if a human being took charge of the task. It is only because the intricacy of the machine is beyond our grasp that we cannot predict the outcome. Not knowing which card will come out of the auto-shuffler is more a matter of our ignorance rather than of pure chance. We think of the machine as dealing cards in a random way because we do not know what it will do. In the same way, when we meet someone 'by chance' we only mean that we had not anticipated the outcome. But an observer, stationed overhead in a helicopter, might have accurately predicted our encounter knowing the direction we were walking.

So it is in the biological world. Variations in gene structures occur, all perfectly conforming to the laws of biochemistry. Yet the complexity of the mechanisms is such that it lies beyond our capacity to predict which particular variation will occur at any one moment in time. And so this is not real chance, only our inability to know and calculate all the relevant factors.

But at the atomic level there is also real chance, and not just ignorance masquerading as chance. Modern physics holds that at this microscopic level we will *never* be able to predict exactly what is going to happen, however good our calculations. Quantum physics, the study of atomic particles, finds that in this micro-micro-world no event is determined in the same way that a dropped hammer will always fall to the ground. Events only have a probability of occurrence. You can never be sure of the when and the what. These quantum probabilities don't apply to our auto-shuffler, for that is dealing out whole playing cards, not minute molecular structures. But in the cellular mechanisms of gene replication it is another story. There will be occasional random errors in the copying of DNA.

Given, then, that the origin of genetic variations may be, at the atomic level, a chancy business, a further level of chance comes into operation. In our card-game illustration, the value of a card to a player depends on the game being played. An ace, especially if it is in the trump suit, is good news for a Bridge player. But the child playing Snap only wants a card of the same value as the one before it. An ace is no more valuable than a three. Unless, that is, the opposing player has just played a matching ace. The rules of the game determine the usefulness or otherwise of the card, and the rules operate completely independently of the shuffling process. You don't shuffle the cards in different ways depending on the game being played. The shuffling is the same for all games. The variations are just dealt to you and, given the rules of the game, you make the best use of them you can.

What is important for our card game is that the shuffler throws up variations with no reference to the game being played, never mind a built-in plan that Mrs Smith should be able to declare 'Five no trumps' and win the game. Similarly, many biologists assume that the gene variations occur with no plan as to their selective advantage in the ecosystem. The gene combinations for being fleet of foot 'just happen', and natural selection encourages the spread of those genes until they become established in the population. In another animal at another time the same gene variation may be of no use whatsoever.

Neo-Darwinism says that at the heart of reproduction and propagation lies this random gene shuffling. Not random in that it breaks the laws of chemistry, but random in the sense

that the consequences for the offspring are not planned beforehand. Without this process of re-shuffling there would be no creative combinations which might allow evolutionary developments. Playing Bridge with a pack that is never shuffled becomes tedious: you know what is coming up. So gene shuffling is as important for a good life-game as card shuffling is for a creative game of cards.

Some never play cards. They say it is all a matter of luck. You either have a good hand or you do not. The whole of card-playing is dismissed because at the heart is a 'random' process. Bridge buffs will retort angrily that this is to miss the important element of skill, and that, given the rules of the game, luck is far from dominant. This is the chance-versus-necessity debate: the balance between chance and law. And we need to examine this through the writings of Jacques Monod.

The Monte Carlo game

'Pure chance, absolutely free but blind, at the very root of the stupendous edifice of evolution,' wrote Jacques Monod in *Chance and Necessity*.

In 1965 Jacques Monod was awarded the Nobel Prize, along with François Jacob and A. Lwoff, for work on genetic mechanisms. But it was five years later, with the publication of *Le Hasard et la Nécessité* that he received widespread public attention. The book, translated into English as *Chance and Necessity*, sought to unravel the mysteries of genetic coding and replication. But it was also a philosophical tract. Monod claimed that the dependence of genetic changes on chance was a sign that accident itself was at the heart of the universe and central to its meaning.

The initial formation of life, Monod claimed, was like playing with a roulette wheel. The outcome was a matter of chance.

'At the present time we have no justification for either asserting or denying that life made only one single appearance on earth, and that, as a consequence, before

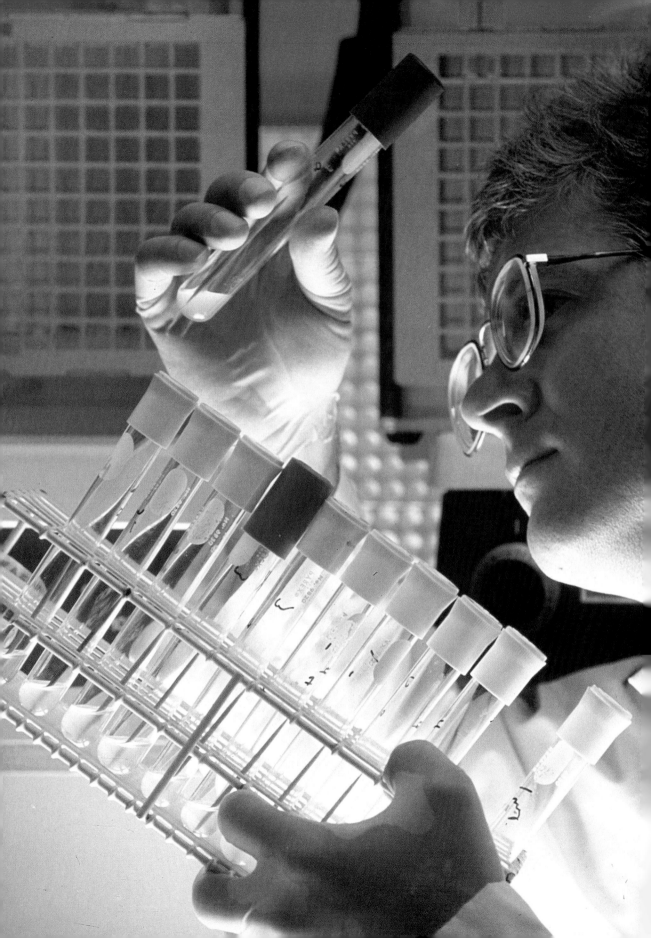

it appeared its chances of occurring were almost nil.

'The universe was not pregnant with life or the biosphere with man. Our number came up in the Monte Carlo game. Is it surprising that, like the person who has just made a million at the casino, we should feel strange and a little unreal?'

We may be glad that we exist. Life is indeed a glorious treasure to have won. But in the combinations and permutations of molecular biology it was simply a chance occurrence, an accident of nature neither foreseen nor planned. 'Man at last knows he is alone in the unfeeling immensity of the universe, out of which he emerged only by chance,' Monod wrote. And, in the same book, 'He lives on the boundary of an alien world; a world that is deaf to his music, and as indifferent to his hopes as it is to his crimes.'

Monod is uncertain whether or not we can face the reality of this creed. All that we find precious, all that we find beautiful, noble or courageous, all must be acknowledged as lucky stopping-points of the roulette wheel. And mankind is at the mercy of the next turn of the wheel, with no divine croupier to stop it at the right moment. In the past, so Monod argues, we have concocted religions and philosophical systems, from Christianity to communism, in order to comfort us in our loneliness. Religion is a human construction to bring us hope in the face of an otherwise meaningless universe.

Yet, however weak our courage, Monod is adamant that we must face the truth. We stand alone, because we *are* an accident of nature. In the past, religions soothed us in that they purported to reveal how we should live. They said that right and wrong were defined outside of ourselves and given to us by the gods. Who now, Monod asks, 'decides what is good and what is evil? All the traditional systems placed

French geneticist and philosopher Jacques Monod believed that blind chance is life's governing principle.

ethics and values beyond man's reach. Values did not belong to him; they were imposed upon him, and he belonged to them. Today he knows that they are his, and his alone.'

In the final paragraphs of *Chance and Necessity* Monod holds out the hope that in choosing to face this austere reality we are in fact choosing to adopt true knowledge rather than the mythical fancies of the past. And only in this choice for 'truth' can there be any inspiration or foundation for other values. In many ways he echoes the elder Huxley, who saw scientific knowledge as the Cinderella hitherto kept in check by the wicked sisters Theology and Philosophy.

What should we make of this bleak, existentialist creed? Where Thomas Huxley understood us to be caught up in the web of unvarying laws, Monod sees the whole fabric of the universe as one gigantic accident.

The practice and teaching of science can never either prove or disprove religious beliefs which deal in ideas by nature not completely open to observation and experiment.

CHANCE OR NECESSITY?

Most of us have at some time thought how nice it would be to win a fortune. But the cards never fall right or the jockey falls off! There is an element of chance in the whole business which makes it improbable or even impossible for us to pick a winner. The Chevalier de Mère, a seventeenth-century French nobleman, was a keen dice player who suddenly hit a losing streak. In an attempt to reverse his bad fortune he wrote to the mathematician Blaise Pascal for help. Pascal in turn corresponded with another famous mathematician, Pierre de Fermat, and between them they set about discovering the laws of chance. In doing so, they established the groundwork of our present-day statistics.

Following Chevalier de Mère, let us find the probability of throwing a six when we roll a dice. The probability of an event is the number of times the event happens divided by the number of attempts. Assuming the dice is evenly balanced, for every six throws we would expect one six. We may throw more than one six, and we may throw none. But there is a probability of one in six that at any one go we will throw a six.

What are the chances of throwing two sixes in a row? The chance of the first dice showing a six is $\frac{1}{6}$, and the chance of this happening with the second dice is the same. And each is independent of the other. On each occasion that the first throw is a six there is only a one-in-six chance that the *next* throw will be a six. So the probability of throwing two sixes in a row is $\frac{1}{6}$ times $\frac{1}{6}$, or one in thirty-six. Probabilities like this have to be multiplied together.

Now, if we apply this sort of thinking to the natural world we meet with some interesting figures. We know that a molecule of water is formed from two atoms of hydrogen bonded to one atom of oxygen. Given these facts there are only three possible structures for the water molecule to take:

H-H-O   O-H-H   H-O-H

Assuming that each of these is equally likely as a structure we would expect a teaspoonful of water to be composed of one-third of each form of molecule. But notice that the first two structures are mirror images of one another and it would not be possible to distinguish between them. So two-thirds of our water sample should be of the form H-H-O / O-H-H and the remaining one-third H-O-H.

If we could take a single molecule from our teaspoonful, what are the chances of it being of the minority form H-O-H? Simple: it is one in three. What is the probability of two consecutive molecules being the H-O-H form? One-third multiplied by one-third gives a probability of one in nine. For three consecutive molecules the chances are $\frac{1}{3}$ x $\frac{1}{3}$ x $\frac{1}{3}$, or one in twenty-seven. The odds quickly multiply and the chance of picking out fifty consecutive molecules of the H-O-H form is one in 717,897,983,000,000,000,000,000! Numbers like this are impossibly large to deal and so they are written in a shorthand notation: $7.17897983 \times 10^{23}$ where the 23 indicates that the decimal point should be placed 23

The answer surely is that we should scrutinize carefully the way Monod has done his philosophy. He is on dangerous ground when he takes the randomness of molecular events as a philosophy for the whole universe. He focuses in on the mechanisms of chance at a genetic level, and then zooms out to claim that everything is chance alone. But what is a chancy business at one level may lead to very definite order at another.

Try pumping up a bicycle tyre. In a bicycle pump the air molecules are circulating with no regard to your attempts to inflate the tyre. As far as you are concerned their movements are random. Yet because there are so many molecules the net effect is that they follow clear-cut laws of compression and expansion. As you push in the pump cylinder the air pressure rises, and some energy is lost as heat. All this a schoolboy could explain. Summed up together the mass of random movements follows clear laws. Randomness at one level does not imply that the final outcome, given thousands or millions of events, is unpredictable as a totality.

Now natural selection is not the same sort of law as that of air compression. The same schoolboy returning to the same bicycle the following month would expect the pump to work in exactly the same way. Not so with the law of natural selection, at least in the real world. In the history of the earth each set of environmental conditions was unique, and so the response to each new gene-combination or mutation was unique. Nevertheless, adding natural selection to random molecular events imposes order and law. The combination of chance modifications

figures to the right.

An average teaspoonful of water contains about 1.67×10^{23} molecules and the probability of each one of those being of the form H-O-H is so infinitesimally small that it is virtually incalculable. A standard home computer will give up the calculation after only eighty molecules! But if we could examine all those molecules we would find that they are indeed of the form H-O-H. Not one is of the alternate form H-H-O / O-H-H!

Are our calculations wrong? Are the laws of chance wrong? No. What is wrong is our assumption that all possible combinations of atoms of oxygen and hydrogen can occur. In fact the structure of the atoms of hydrogen and oxygen is such that they can only combine in one way. They are constrained by the laws of chemistry and physics to form H-O-H molecules and no others.

The spontaneous formation of patterns within non-living systems is well known — for example, snowflakes and some forms of polymer gels which absorb water and swell to form hexagonal patterns. These regular patterns are the result of natural laws and they are not formed under the control of genes as in animals or plants. But the structures we see in plants and animals are made from atoms and elements just like the structures in the inanimate world. As such they are subject to the same laws of chemistry and physics as the rest of the world. A drop of ink falling through a beaker of water can adopt a shape which is very similar to that of a jellyfish. If in the case of the ink drop the shape is determined by the laws of physics, can we be justified in ignoring the importance of such laws in relation to living organisms? How was the jellyfish itself formed?

A number of scientists feel that it is not unreasonable to assume that the forms of living organisms are also constrained by these physico-chemical laws and that they do not have unlimited freedom so that their parts combine in a purely random way. Certain forms are therefore more likely to develop than others. Chance may affect which molecule combines with which, but necessity dictates how they will join together.

To a neo-Darwinist an organism is made up of a large number of individual traits, each controlled by a gene. The form of any individual organism is the accumulation, over a long period of time, of small variations caused by random mutations of these genes. As all mutations are equally likely, the number of different forms possible is astronomically large. Natural selection weeds out the poorly adapted forms and leaves behind the well adapted ones. But it would seem that some organic structures are more likely to be produced than others in response to physical and chemical laws. And so plants and animals are not solely the result of chance events at an atomic or molecular level in the genes. They are the product of the necessity of following the physical and chemical laws that govern the rest of the world.

plus the filtering effect of natural selection produces a genuinely creative system that allows new and novel properties to emerge. In our game of cards the combination of luck and the rules of the game produces an endless series of hands, each different from the other. And yet, because of the rules, each hand moves towards certain ordered structures, such as a collection of 'sequences' or 'runs'. The details of a winning hand are unknown beforehand, but the nature of the game dictates that there will be a winner.

Now we need to step even further back. Who creates the cards such that they can be readily shuffled? Cards which are of uneven size, shape or thickness make shuffling very difficult. It is a simple point, but the manufacturers of cards design them for easy shuffling. They expect you to enjoy the luck element in any card game. And then who writes the rules of the game? Is there a reason why the structure of the universe is such that matter allows these chance alterations, and that the combination of gene shuffling and the law of natural selection offers such a marvellous agency for creating variety? The believer in God need not shrink back from recognizing that chance events are at the centre of the universe. It is a superb system for exploring endless potentialities within nature, and the way the sovereign God works through nature as Creator.

Jumbo jets and junkyards

The intricate workings of the human eye make you marvel. For Archdeacon Paley it proved beyond doubt the existence of a designer behind the universe. The eye is like a telescope, but infinitely more complex. If the simple telescope is designed by human hand so, by analogy, must the eye have been designed by the Almighty. So reasoned Paley, but Darwin refuted this argument, claiming that the apparent design in nature is a product of evolutionary forces operating over millennia. Nevertheless the wonder of the eye still amazed him, and troubled him.

Returning briefly to our Bridge game. Given that the cards are well shuffled, how do you react when you are dealt a hand, say, of all thirteen trump cards: the ace of hearts right through to the king? Since the chances of this happening are slim (when did it last happen to you?) you may question the quality of the shuffling. Certainly your opponents will if you are the dealer! Surely the hand has been rigged?

There are many who doubt the power of natural selection to accomplish such feats of engineering as the human eye. It seems to them impossible that something so intricate, so sensitive and beautiful as the eye could have been thrown together over the vast reaches of time by the steady drip of genetic changes. It is a favoured argument of creationists to calculate the statistical probability of the molecules necessary for life (say the formation of DNA) arising spontaneously in a primeval biological soup. The chances are next to zero. They are about as probable as the pieces of scrap in a junkyard being blown together on a windy day to form a jumbo jet.

But they misunderstand the process of evolution by natural selection. And to clear up the misunderstanding we must return once again to our game of cards. This time we shall need to play 'Rummy' or 'Canasta' — the sort of game where the aim is to collect sequences of cards or groups of the same value. In any such game the chances of being dealt a winning hand right at the start are small (though not nearly as small as for the spontaneous formation of DNA). You could spend many hours dealing yourself hand after hand before a winning

combination arose. But the aim is to get a winning sequence and, as the game progresses, each player discards cards from his or her hand and picks up new ones until the required sets are formed. The stack of cards in the centre is a source of new (and random) cards. Each player throws away those cards of no benefit to the combinations he or she is collecting, and retains others. False leads are often started — a decision to collect kings may be regretted later when an opponent turns out to have the same strategy.

Natural selection also progresses by the same cumulative process. The chances of the eye being formed over one biological generation by a chance re-ordering of genes are vanishingly small. But the chances of a slight change occurring on the step towards an eye, such as a light-sensitive spot which give some selective advantage to its owner, are higher. In the card game, new cards are either picked up, then added to the hand if they are useful or discarded if they are not. In evolution, new forms which have an advantage, however small, survive, while others without the new genetic make-up are gradually overtaken. There is nothing conscious about this, of course; no deliberate collecting of 'cards'. Darwin was often criticized because he made natural selection sound like a divine Mother Nature hurrying through her flora and fauna picking the best with which to seed the next generation — a cosmic gardener tending the roses and pulling out the weeds. But this was merely an incautious mode of expression: Darwin certainly meant that natural selection was unconscious selection. Evolutionary changes only accumulate because less fit variations die out or breed in fewer numbers.

The British biologist Richard Dawkins describes a possible sequence of small steps gradually moving towards a complete eye:

'Some single-celled animals have a light-sensitive spot with a little pigment screen behind it. The screen shields it from the light coming from one direction, which gives it some 'idea' of where the light is coming from. Among many-celled animals, various types of worm and some shellfish have a similar arrangement, but the pigment-backed light-sensitive cells are set in a little cup. This gives slightly better direction-finding capability ... and if you make a cup very deep and turn the sides over, you eventually make a lensless pinhole camera. A pinhole camera forms a definite image, the smaller the pinhole the sharper (but dimmer) the image, the larger the pinhole the brighter (but fuzzier) the image. The swimming mollusc *Nautilus* has a pair of pinhole cameras for eyes ... '

According to the Darwinians, all you need is a sequence of small steps. In some instances the biological world is scattered with the previous 'hands', held by organisms which have not progressed beyond them. The *Nautilus* would benefit from a lens (much as its relatives the squids and octopuses do) yet it manages without. However, in many instances in the biological world the final product has so superseded the former versions that they no longer survive. We have no idea what were the precursors to DNA.

The German scientist Manfred Eigen believes that even among the raw chemicals of life chance may not be so central to evolutionary development as previously thought. As a

Nobel prize-winning chemist Manfred Eigen believes that a kind of natural selection operates even at the level of molecules.

The aquatic creature 'Nautilus' has pinhole cameras for eyes. Nature has examples of creatures whose evolutionary development stopped at what seems like an intermediate point.

chemist who won his Nobel Prize for finding out just how rapidly chemical reactions proceed among the working molecules of life, Eigen is clear on the difficulties of assembling a particular sequence of chemical units simply by chance in a given time. Yet because of internal chemical restrictions (certain chemical bonds are easier to form than others) and because certain intermediate structures are more stable than others, there may well have been a sort of 'natural selection' occurring even at the molecular level. The result is that quite complex chemical structures necessary for life are far more readily formed than chance alone would suggest. The odds against the sudden, yet complete, formation of a DNA molecule are

181

greater than the atoms in the universe. Yet in the early days of the earth's history, simple building blocks may have occurred not just once but many times. The route from these sub-units to the final DNA is still shrouded in mystery. But ignorance is insufficient reason for us to plead impossibility.

In his *Laws of the Game*, 1975, Eigen writes:

> 'Natural law means a channelling, if not a taming of chance. We human beings are just as much the product of this law as of historical chance.'

We may have great difficulty in determining the intermediate steps along any evolutionary path, but there is never a claim that the final product arrived all in one go. Life is not 'chancy' in that way.

The Blind Watchmaker

To demonstrate the awesome ability of natural selection to build complex structures, Richard Dawkins has written a book entitled *The Blind Watchmaker* (1986). He takes his title from the work of Paley, whose analogy of creation needing a designer much as a watch needs a watchmaker caught the imagination of Darwin and most of Victorian Britain. For Dawkins, however, the analogy is false. Paley may have argued with passionate sincerity, informed by the best scholarship of his day. But his analogy was 'wrong, gloriously and utterly wrong'.

> 'The analogy between telescope and eye, between watch and living organism, is false. A true watchmaker has foresight: he designs his cogs and springs, and plans their interconnections, with a future purpose in his mind's eye. Natural selection, the blind, unconscious, automatic process which Darwin discovered . . . has no purpose in mind. It has no mind and no mind's eye. It does not plan for the future. It has no vision, no foresight, no sight at all. If it can be said to play the role of watchmaker in nature, it is the *blind* watchmaker.'

Dawkins argues that natural selection is a sufficient explanation for all phenomena. We saw above how he begins to explain the origin of the human eye. Within the life-game the shuffling of the genes throws up variations that have an advantage (however small) over those that had gone before. It is these advantaged forms which propagate and survive in greater numbers — and so the march of evolution continues. Dawkins does not (and cannot) know whether in fact everything did evolve by means of his all-powerful natural selection. His example of the development of the eye is only a picture of what might have happened. He believes, as an act of personal commitment, that cumulative selection can — and did — accomplish everything without God. His trust is as deep as Thomas Huxley's in the all-embracing web of natural causes. One hundred years might have elapsed yet the faith is the same.

Where in this scheme is the place for God? If variations plus natural selection are all-sufficient, why should we invoke a creator-designer?

Dawkins sees no reason at all:

> 'Many theologians who call themselves evolutionists . . . smuggle God in by the back door: they allow him some sort of supervisory role over the course that evolution has taken, either influencing key moments in evolutionary history, or even meddling more comprehensively in the day-to-day events that add up to evolutionary change.'

He recognizes that he cannot disprove such a role for God: he simply sees it as superfluous. Others are not so readily convinced. Evolution by natural selection may well be the mechanism by which creation was accomplished, but an understanding of the mechanism still leaves unanswered the deeper questions of why the universe and human consciousness exist at all. Dawkins' explanations may be accurate as far as they go. But they do not go far enough. They are like descriptions of a recording of a Bach fugue that might be given by a studio engineer. The engineer rapturously explains the source of the audio tones, their varying amplitude and fidelity of reproduction. But Bach himself is left out of the picture.

In an earlier book, *The Selfish Gene*, Dawkins claimed that we humans are simply advanced machines for the propagation (or, using his terms, replication) of genetic material:

'The replicators which survived were the ones which built *survival machines* for themselves to live in. The first survival machines probably consisted of nothing more than a protective coat. But making a living got steadily harder as new arrivals arose with better and more effective survival machines. Survival machines got bigger and more elaborate, and the process was cumulative and progressive.

'Four thousand million years on, what was to be the fate of the ancient replicators? They did not die out, for they are past masters of the survival arts. But do not look for them floating loose in the sea; they gave up that cavalier freedom long ago. Now they swarm in huge colonies, safe inside gigantic lumbering robots ... They are in you and in me; they created us, body and mind; and their preservation is the ultimate rationale for our existence. They have come a long

way, those replicators. Now they go by the name of genes, and we are their survival machines.'

Now this is the language of hyperbole, but even so Dawkins is taking us back to the unyielding despair of Bertrand Russell. To Russell's 'accidental collocation of atoms' Dawkins adds a further epithet (or should we say epitaph?) for human existence: human beings are the Cadillac or Rolls Royce of gene survival machines.

Those of us who believe in a God behind the cosmos do so not because we can find no other way to explain the mechanisms of the universe, whether it be the movement of planets or the evolution of people. Our belief comes ultimately from reflecting on the very existence of laws, beauty, morality and human consciousness. What does it say about our cosmos if all these things naturally arise?

We are not contending, as Paley did, that the design in the universe proves the existence of a Designer. We cannot do this for the same reasons that atheists cannot use Darwin's theory to disprove God's existence. A discussion at the level of mechanism will never fathom the deeper reaches of authorship just as a talk with a television engineer on how to adjust the picture will never reveal details of the programme's scriptwriters. Dawkins is satisfied

---

THE MORALITY OF SELFISH GENES

Richard Dawkins in *The Selfish Gene* believes Tennyson's famous phrase 'nature red in tooth and claw' sums up our modern understanding of natural selection admirably. But he is not prepared to let this determine his code of morality. His biology only allows him to say how things have evolved and not how humans morally ought to

behave. 'My own feeling,' he writes, 'is that a human society based simply on the gene's law of universal ruthless selfishness would be a very nasty society in which to live.'

Here the sociobiologist is in a quandary. Dawkins encourages us to teach generosity and altruism yet spends the remainder of his book showing us how all our behaviour derives from our genetic make-up. Only mankind, he avers, can upset the designs of the selfish genes. Only humanity can rise above the programming of the past. Dawkins faces the same difficulty as Thomas Huxley one hundred years before. Evolution determines

our nature, yet the implications of this are so awful that, at the same time, we must be allowed freedom to transcend our lowly origins. Dawkins believes it is a common fallacy 'to suppose that genetically inherited traits are by definition fixed and unmodifiable.' He says we are not necessarily compelled to obey them all our lives.

Dawkins is entitled to this act of faith, but he has very little on which to base it, save his own horror of jostling genes becoming jingoistic generals.

THE GHOST IN THE MACHINE

Julian Huxley pointed to the progressive nature of evolution. From our human perspective there does seem to have been an ever-increasing development of complexity and consciousness.

To explain this sense of direction within evolution a number of biologists in the first half of the century believed that working alongside the laws of gene mutation and natural selection was a further life force. The known laws of science were not enough, there had to be something extra still undiscovered. Here Henri Bergson, who wrote *Creative Evolution* in 1920, was perhaps the most influential thinker, although his *élan vital* was only a creative force, leaving the resulting direction to chance. The term 'orthogenesis' was given to theories which proposed a determined direction for evolution, and the title 'vitalist' given to biologists who believed in some vital force pushing creation ever onwards. Vitalism is now regarded as an embarrassment. No one could isolate this extra force. And although many aspects of evolution are still a mystery no one feels a need to propose unknown forces. But at the time vitalist views were seen as holding out a hand to theology, for God's Spirit could be readily equated with the life force.

But one further puzzle remained. What is the origin of consciousness and our sense of freedom? If everything was explainable in terms of scientific laws, whence our capacity to think and to feel free?

In the 1920s and 30s philosophers such as A.N. Whitehead took our awareness of human freedom and consciousness as axiomatic for the whole of creation. Nothing is simply material stuff, all has a 'mental pole' as well as a 'physical pole'. All matter has the potential to display mental activity, although the degree of such activity reduces sharply as we descend from animals to inanimate objects. When adapted to a Christian perspective (by, for example, Charles Birch) God's relationship with the cosmos is seen not so much as determining each and every event, from a radioactive element's decay to the sinking of the *Titanic*, but of willing and encouraging matter along its course. Like a piano teacher goading her pupil, matter is endued with evolutionary potential, and it is God who draws it out. God does not determine all events, but he does influence them.

Freedom lies in the very structure of the universe. This concept has the advantage of answering Darwin's worry over nature 'red in tooth and claw'. God is no longer seen as strictly ordering each and every mutation, especially those which are harmful.

But is all this true to reality? Ascribing to all things some level of mental activity seems strange. We are brought up firmly to believe that material objects are somehow fundamentally different from living organisms. And probably we are right. More recent thinking no longer sees the necessity to project right back along our evolutionary path the consciousness we now enjoy. As matter becomes more complex so new (and quite distinct) properties emerge. So human consciousness creatively emerged somewhere *en route*. There is no need to suggest it was always there, in some very limited way, among the first molecules.

But if there is no consciousness, however limited, within matter how does this property arise from nothing? There has been a long tradition of viewing 'spirit' as something other than matter. It is something given when he has explained how cause leads to effect, and assumes he tells us the whole story. He is like the electronic engineer in relation to the Bach fugue. Bach is left out. In the same way scientists, and Dawkins recognizes this, have no authority to say that their explanations in terms of mechanisms are the only possible explanations. We are entitled to break out of this self-imposed prison and see the mechanism as having an author. Indeed for many this is a far more realistic view of the world. It makes more sense of the cosmos.

The origin of the words on this page may be expressed in terms of printer's ink and wood pulp. You could trace the movement of letters from a word processor to the printer's films, and then from the films to the printed page. The words move from one 'survival machine' to another and the description is complete — in scientific terms. Yet to leave out of account our intention as authors, to fail to see the meaning behind the ordered sequences of letters, is to miss the point. Well, we think so. Dawkins has given us an inspired description of the power of evolution by natural selection. But to many of his fellow survival machines he appears to be blind to the Maker who watches over us.

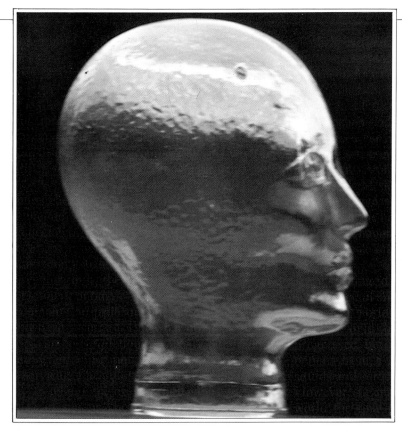

products of brain cells working according to the laws of chemistry.

A middle option is available between reducing everything to chemistry and believing in an extra, yet non-material, spirit. This viewpoint says that consciousness emerges as a totally new and creative property of matter, given sufficient complexity. The new phenomenon cannot be explained simply in terms of physics or chemistry. It is on a different level with new properties needing different descriptions.

Take the letters of the alphabet. Suppose you are limited to using words of only two or three letters. Conversation would be limited, to say the least! Allowing four or five letters would vastly increase your vocabulary and further increases would quite suddenly open up the range and beauty of language. Now language is more than simply letters strung together, although it may be expressed in letters. It operates on a different level. Similarly mind and consciousness arise with increasing biological complexity, and take on properties of their own. We may study the electrical brain responses that accompany thought, but we cannot reduce Plato to nerve pulses, or Aristotle to alphawaves.

supremely, and many would say uniquely, to us humans. Descartes thought all animals were mere complex machines — it was only mankind who possessed a soul. Somehow soul and body exist independently, side by side.

When it became impossible to investigate this non-material element of creation many dismissed it. They came to think that only matter was real. And so our deepest thoughts were reduced to nothing but the

God and his creation

If we do believe that there is a Maker, how should we understand his relationship to his creation? Three approaches have commonly been discerned:

☐ *The least satisfactory is 'deism'.* On this view God creates the world as an engineer makes a machine. Then, having wound it up and set it going, he retreats to watch his handiwork. The world is autonomous; it is sustained by the power invested in it at creation.

Deism is unsatisfactory because it does not match up to the God revealed in human history

and the pages of the Bible. We do not have a creator who has simply left us to get on with it. The world is not a gigantic firework of which God has ignited the touch-paper and retreated to a safe distance!

☐ *Then there is a kind of 'semi-deism',* which sees an ongoing role for God in adjusting and advancing the creation he has made. Having first created the world we find that at certain crucial moments — the formation of life from inert matter, the creation of humanity — God once more steps in with miraculous power. God is most evident wherever there are gaps in ordinary explanation; what we cannot explain by the operation of natural cause and effect must be the miraculous power of God.

Semi-deism fast approaches deism as science uncovers more and more natural explanations for phenomena. The lightning strike is no longer seen as the wrath of God, but as a sudden discharge of electrical energy. And only a child now refers to thunder as 'God moving his furniture'! As more and more mysteries are explained, the space left for God is for ever shrinking. And even if a healthy supply of unsolved questions still remains, we are left with doubts as to what God does when there is no need for a special intervention. The occasional miracles by God might assure us of his power and benevolence, but they also speak of his more regular absence, like a landlord who only shows up when new tenants are required.

As an Oxford clergyman, Aubrey Moore, wrote back in 1889:

'The one absolutely impossible conception of God is that which represents him as an occasional visitor. Science has pushed the deist's God further and further away, and at the moment when it seemed as if he would be thrust out altogether, Darwinism appeared, and, under the disguise of a foe, did the work of a friend. It has conferred upon philosophy and religion an inestimable benefit, by shewing us that we may choose between two alternatives. Either God is everywhere present in nature, or he is nowhere.'

☐ *But a third, traditional understanding, often named 'theism', has seen God as working continually through the nexus of events.* The scientific laws we observe are in reality descriptions of the regular working of God. Miracles are possible, but they are not extra interventions. God is at work all the time — a miracle is only a different (and irregular) mode of action.

This view rightly stresses the continuing presence of God within creation. The English biochemist and theologian, Dr Arthur Peacocke, expresses the relation of God to his world as similar to the relationship between a person's thoughts and his actions. The physiologist can describe the moving of an arm in terms of moving muscles, chemical messengers and the like. The person himself relates the movement to his thoughts and intentions, whether he was raising his arm to strike or to embrace. There is a correspondence between the mental and the physical, yet we sense that the real initiator of action is the conscious person, not the mechanisms of the body. Our thoughts and intentions control our actions. Descriptions of physical movements will never reveal their meaning. Biology can only examine the interlocking world of neurons and synapses.

Lest we get into a muddle over which comes first, the person's thoughts or the brain waves expressing those thoughts, it must be pointed out that Peacocke is only using this human paradox as a picture. Every day we know the reality of having wishes and intentions, and these are translated into bodily actions. So, in a similar way, the physical processes in the world may be pictured as the outcome of God's will and intention. The physicist, the chemist, the biologist will never directly see this. They will only ever see the regularity of laws. But the believer holds that behind these laws lies the mind of God.

The picture of thought and action tends to ignore any understanding of human freedom. Action flows directly from thought, and without thought there is no action. But if the physical world is an outcome of God's will, does this mean that everything (including ourselves) is controlled from above like a puppet? Can we not act independently of God? Peacocke reminds us that many bodily functions regulate themselves — for instance, our breathing, our body temperature and balance. Perhaps, then, there are pockets of activity within God's world which can act independently of him.

It is one of the marvels of creation, perhaps the chief marvel, that God has given his creatures a measure of freedom. Within his overarching purposes he devolves freedom on us. No human analogy is readily to hand since such a gift of freedom is not within the capacity of human inventors to design into their own artefacts or machines.

Some Christian theologians would go further and see the chance processes in the world (the mutation of a gene, the decay of a radioactive element) as pockets of independent action established by God. Such chance processes were designed by God as a method of exploring all the potentialities of creation. Others would disagree, pointing out that what is perhaps 'chance' for us may well be known by God. He it is who designed the shuffling mechanisms and the interacting rules of the life-game. We may not know the outcome, but he assuredly does.

Inevitably in talking of the ways of God we are reduced to using paltry human pictures and analogies. Some may reject the reality through the inadequacy of the explanation. Deep doctrines, such as that God is one person in three, cannot be readily grasped by us. Our understanding is finite. Yet we can know the reality of God, Father, Son and Spirit in our daily experience. So the understanding of a creator God working through the regular laws of nature may be difficult to picture. In the end it is an article of faith. Yet because of their experience of God many find it a faith which makes the most sense of the mysteries of the cosmos.

A statement of faith

The Christian understanding that God is all the time working through what we term 'natural laws' is a statement of faith. Richard Dawkins says such a faith is superfluous. He believes that the processes by which mankind has evolved do not *demand* a religious explanation, only genetic mutations, natural selection and bags of time.

We are inclined to agree. Within all spheres of science it is possible to provide explanations simply in terms of the causes and effects at work. To be a good scientist you do not need to use God as an extra force somehow fitted in among the others. God is not a God-of-the-gaps. He is not a mysterious source of power invoked whenever the regular scientific disciplines cannot find an answer. As science extends its sway this sort of God becomes vanishingly small. No, God is behind, and within, and working through the natural phenomena, and a descriptive subject such as science will only ever be able to watch the interlocking network of his actions, and never discover the author himself.

The writer Dorothy L. Sayers once used the telling example of a play performed on a stage. You may study the plot and admire the stage directions; you may marvel at the set and applaud the actors. But you will never meet the playwright. Unless, that is, he or she makes a personal appearance in the play, or takes a bow at the final curtain. So science, like a drama critic, can only describe the play of forces against each other. God, as author, watches unseen while his own creation, moment by moment, unfurls on the stage.

For religious believers to adopt this view will seem to many to betray a weakness in their belief. If God cannot be measured and proved then why should we believe? The creationists seek to respond to such objections by giving God a role in creation that is clear for all to see, pointing to miraculous interventions in forming the earth. They fear that the apparently weak doctrine of God working through processes such as natural selection pushes God too far into the background, so that our moral and social life is threatened. And recent history suggests that their fears are well-founded. Evolutionary biology has spawned the philosophy of evolutionary humanism, while the Huxleys of our society seek to melt down the traditional gods and erect new, scientific ones in their place. Monod and Bertrand Russell want to rid us of religious belief altogether. In a way creationists look to a God who makes material-istic interventions to persuade people who have problems reaching beyond the material world.

But this is not the Bible's approach. The Bible writers do not argue from the creation stories to a belief in God. The focus in both the Old and New Testaments is on the God of human history, not just the God behind science.

It sees God as a player within his own play as well as the author of the whole drama. Just as the opening hymn of a church service sets the tone of worship but is not the keypoint of worship, so the creation account reveals God as the author of the drama but is not itself the centre of the story. The reality was (and is) the working of God in human lives and in the whole sweep of history.

For the Old Testament the dominant motif is God choosing the people of Israel, freeing them from slavery in Egypt and forming a ragged tribe into a special people. He spoke to them through Moses and the prophets. The Israelites marvelled at their national history, but only as they reflected on their history did they see that their God was indeed the God of all people. They came to recognize that God was in control not only of Israel's story but of the destiny of the neighbouring tribes as well. He ruled over the worthless idols of Canaan or Babylon; he ordered the forces of nature and banished chaos from the world. The first two chapters of Genesis are the supreme expression that God is Lord of the universe. But the Israelites came to understand God's universal sovereignty through what he did for them as a people rather than simply because he had created them.

Similarly in the New Testament the focus is on the Son of God coming into our human experience. It is his teaching, his death, his resurrection which form the kernel. Only because the early missionary Paul had met the risen Christ could he stand in the Greek marketplaces and declare that their creator, to them unknown, had at last been revealed. His own experience led to the recognition of God's cosmic sovereignty in Christ.

The centre of Christian worship is the Holy Communion or Lord's Supper. Here the church remembers what God did for us historically in Jesus and the spiritual presence today of Jesus Christ among us. Here is the reality of faith. Yes, we have our harvest services, when we thank God for the regular laws of the created world. We sing our opening hymns about 'All things bright and beautiful'. But we move to our understanding of the God of the cosmos because we have seen him in history, we have found him in ourselves. We believe in the reality of God because he has been revealed in Jesus. And that same Jesus miraculously rose from the dead and lives among his people today by his Spirit. It is because the Christian faith is so centred on the living presence of God that we do not need to look to science for 'proof' of his existence. Unless there is overwhelming evidence in favour, we do not need to show that God worked by special miracles at creation. Most people believe in Jesus not because they first believed that God created the world, but rather the reverse. For faith, the cross and the empty tomb are eloquent enough.

The Bible itself contains no proof for God: his existence is everywhere assumed. But what is discussed, from Genesis to Revelation, is our response to him and whether or not we will be faithful in our service. The apostle Paul knew our readiness to claim to be so wise and yet . . . and yet to exchange 'the glory of the immortal God for images resembling mortal man or birds or animals or reptiles'. We so readily turn to other 'gods'. We prefer the myths of creation to the truth of Genesis.

How was Genesis heard in its own day? Conrad Hyers has written:

> 'When one looks at the myths of surrounding cultures (then threatening Israel) one senses that the current debate over creationism would have seemed very strange, if not unintelligible, to the writers and readers of Genesis. What pressed on Jewish faith from all sides, and even from within, were the *religious* problems of idolatry and syncretism. The critical question in the creation account of Genesis 1 was polytheism versus monotheism. *That* was the burning issue of the day, not some issue which certain Americans 2,500 years later in the midst of a scientific age might imagine it was.'

For this reason it is so mistaken to treat

the first chapter of Genesis as science. It is a literary statement of the universal lordship of God, and mankind's utter dependence on him. It is a story of the wonder of our creation, yet the awful readiness of our rebellion. Israel's neighbouring tribes had no experience of God leading them out of Egypt. For them the sun and the moon, light and darkness, watery chaos and fertile land, vegetation and animals, these were their gods: each to be feared and worshipped, each to be remembered in an annual ritual which maintained their place. Belief in this pagan cosmogony was the threat to Israel's life. And so the writer of Genesis performs a neat hatchet job in ascribing all these so-called deities, one by one, to the power of God the Lord.

And what of today? A reading of history which pits the church against science is too simplistic. Indeed it is wrong. That myth was cradled by the advocates of emerging secular faiths as they jostled for cultural supremacy. If we are to employ battle imagery then the real warfare is between Christianity and a humanistic or atheistic view of the world. The church takes her stand against the inroads of chaos and the twentieth-century gods of Progress and a materialistic world-view.

Genesis then rings as true as ever, whether one follows an evolutionary account of biological origins or not. The phrases of this majestic first chapter still insist that God is Lord and that mankind is not under the sway of capricious elemental forces, or without hope or meaning in life. We have no need of other gods or the manufacture of creation myths.

We need not worship with Thomas Huxley at the altar of all-pervading natural law, nor derive our security from the march of evolutionary history.

We need not believe we are random collocations of atoms, nor swallow the lie that life is a biological accident.

We need not accept our lowly status as genetic survival machines, nor fear that one day we will be discarded in favour of more advanced models.

Instead we can apply the poetry of Isaiah as we discard these modern worthless images. Our number did not just come up in the giant Monte Carlo game of life, but in the providential heart of God.

> But when I look there is no one; among these there is no counsellor who, when I ask, gives an answer.
>
> Behold, they are all a delusion; their works are nothing; their molten images are empty wind.
>
> "To whom then will you compare me? Or who is my equal?" says the Holy One.
>
> Lift up your eyes and look to the heavens: Who created all these?
>
> Do you not know? Have you not heard? The Lord is the everlasting God, the creator of the ends of the earth.

INDEX

ILLUSTRATIONS

Aip Niels Bohr Library, Copenhagen 140
BBC Hulton Picture Library 25, 55, 60 (top), 102, 103 (right), 112 (top), 127, 128, 135 (bottom right and top left), 156, 172
The Bridgeman Art Library 30/31, 49, 57, 59 (top)
Cambridge University Library 73 (right)
Camera Press 143, 181 (top)
Tony Cantale Graphics 62/63
Mary Evans Picture Library 11, 14 (top), 21, 26, 27, 29, 37, 47, 48, 53 (bottom right), 60 (bottom), 70, 71, 76 (bottom), 77 (left and right), 79, 80, 83, 112 (bottom), 117 (top), 119, 123, 174, 177
Fotomas Index 63 (top inset)
Historical Newspaper Service 135
Michael Holford 129
D. Horwell 64, 66, 67
Illustrated London News 68, 121
Jenny Karrach 53 (top), 113
Jill Krementz 145
Linnaeus Society 18, 19 (top and bottom), 44
Lion Publishing/V. Blackmore 27, 42, 52, 53 (bottom left), 58, 59 (bottom), 85, 101, 106 (top and bottom), 107, 114, 123, 165 (top), 168 /D. Townsend 24 /J. Willcocks 185
The Mansell Collection 13, 14 (bottom), 35, 40, 61 (bottom), 73 (left), 75, 76 (top), 81, 103 (left), 109, 120, 126, 135 (inset, top right), 165 (bottom)
National History Photographic Agency 9, 17 (right), 23, 51, 92, 95, 144, 181 (bottom)
Natural History Museum 84
Michael Poole 76 (centre), 78
Popperfoto 155, 171
Royal Botanical Gardens, Kew 115 (Inset, and bottom)
Royal Observatory Edinburgh 50
Science Photo Library 137, 139, 141
Sunday Times 161
Swedish Linnaeus Society 20
University College London 43
Victoria & Albert Museum 162
ZEFA (UK) Ltd 17 (top), 23, 32, 33, 34, 39, 61 (centre and top), 65, 69, 88, 89, 90, 91, 97, 116, 117, 148, 151, 167, 173, 176
Zoological Society 12 (inset and bottom)

QUOTATIONS

Richard Dawkins, *The Blind Watchmaker*, Longman Scientific & Technical, 1986, pages 5, 85, 316
Conrad Hyers, in *Is God a Creationist?* (ed. R. M. Frye), Charles Scribner's Sons, 1983, page 100
Richard Leakey, *The Making of Mankind*, Book Club Association, 1981, pages 245-47
Carl Sagan, *Cosmos*, MacDonald Futura, 1980, pages 337-38